Oil and Gas Processing Equipment

Equipment
Risk Assessment with Bayesian Networks

Oil and Gas Processing Equipment

Equipment
Risk Assessment with Bayesian Networks

G. Unnikrishnan

CRC Press
Taylor & Francis Group
Boca Raton London New York

CRC Press is an imprint of the
Taylor & Francis Group, an **informa** business

First edition published 2021
by CRC Press
6000 Broken Sound Parkway NW, Suite 300, Boca Raton, FL 33487-2742

and by CRC Press
2 Park Square, Milton Park, Abingdon, Oxon, OX14 4RN

© 2021 Taylor & Francis Group, LLC

CRC Press is an imprint of Taylor & Francis Group, LLC

Library of Congress Cataloging-in-Publication Data
Names: Unnikrishnan, Ji., 1944– author.
Title: Oil and gas processing equipment : risk assessment with Bayesian
networks / authored by G. Unnikrishnan.
Description: First edition. | Boca Raton : CRC Press, 2021. | Includes
bibliographical references and index.
Identifiers: LCCN 2020019126 | ISBN 9780367254407 (hardback) |
ISBN 9780429287800 (ebook)
Subjects: LCSH: Gas manufacture and works—Risk assessment—Mathematics. |
Petroleum refineries—Risk assessment—Mathematics. | Gas manufacture
and works—Equipment and supplies—Safety measures—Mathematics. |
Petroleum refineries—Equipment and supplies—Safety
measures—Mathematics. | Bayesian statistical decision theory.
Classification: LCC TP752 .U66 2021 | DDC 620.1/07—dc23
LC record available at https://lccn.loc.gov/2020019126

ISBN: 978-0-367-25440-7 (hbk)
ISBN: 978-0-367-54197-2 (pbk)
ISBN: 978-0-429-28780-0 (ebk)

Typeset in Palatino
by codeMantra

This book is dedicated to my parents and wife

For a safer world

Contents

Preface

Quantitative risk assessment (QRA) is embedded in the design and operation of oil and gas projects and facilities. This practice has matured over the years. Theories, calculation techniques and software are available for computing the key components of QRA, namely the release rates, frequencies, consequences and risk. The results of QRA are primarily used for decisions on land use planning.

However, when an accident happens, the findings often point to several layers of causes lined up in a rather unusual coincidence. Whether it is Jaipur tank farm or Buncefield, Texas City refinery, the story is the same. The striking fact is that, these were all avoidable! From my years of experience in the design of oil and gas facilities and in conducting/reviewing several quantitative risk assessments/reports, I found that conventional QRAs are limited in its ability to predict such situations. The above aspect together with my research on Bayesian networks (BN) gave me the idea of writing such a book.

Bayes theorem, its developments and application in many areas, provides a sound basis for logical reasoning of cause and effect. Representation of causal mechanisms through Bayesian network opens up the potential for analyzing the layers of causes in a complex facility that can eventually line up and result in an accident. Though BNs are being applied in diverse areas such Computer Science, Finance, Ecology, Medicine, etc. it is not being applied to a great extent in oil and gas industries.

The book aims to introduce such applications and avoids the complex theories. The contents have been kept to a basic level. I hope the book will be beneficial to persons in academic and industrial field.

The book does not claim to be error free. There is scope for improvement. Any feedback is welcome.

I appreciate the support and encouragement given to me by my PhD thesis advisors Dr Shrihari and Dr Nihal Siddiqui at the University of Petroleum and Energy Studies, Dehradun, India during my research on the subject. Writing of a book takes time and effort. Usually, it is the spouse who sacrifices her holidays and enjoyments to make it possible. Here also it is not different. I thank Dr. Gangandeep Singh of CRC Press for his persistence and help in finalizing the draft. I also acknowledge and thank the project managers Ms. Sofia Buono & Ms. Rebecca Dunn at CodeMantra for the careful attention and collaboration in production of the book. I thank my wife Girija for her patience while I worked on the book.

G. Unnikrishnan PhD.

Author

G. Unnikrishnan has over 40 years of experience in oil and gas industry. His experience spans the areas of process design, process safety, engineering & project management. He is currently on an assignment as engineering specialist with a National Oil Company in the Middle East. He previously worked with engineering consultancy companies in India and abroad. His current work involves review, assessment and providing advice on Front End Engineering Design and engineering management for upstream oil and gas projects.

He is keenly interested in optimization of process design and how it can be done with the highest process safety. He believes that much needs to be done in process plant design and operations to minimize accidents. He is an active researcher in the area and has presented and published papers on the subject in several international conferences and technical journals.

He is a certified Functional Safety Engineer on Safety Instrumented Systems. He holds a degree in chemical engineering from Calicut University, MTech from the Cochin University of Science & Technology and PhD from the University of Petroleum and Energy Studies, Dehradun, India. He can be contacted at ukrishnan77@yahoo.com.

1

Introduction

The oil and gas industry handle highly inflammable and toxic fluids under severe operating conditions and have high inventories of the same. Given the hazardous nature of its operations, it is important that the industry ensures such facilities are designed, maintained and operated in a safe manner. Different methods have evolved over time to analyze and mitigate the risks involved. However, major accidents continue to occur and at that time issues on safety and risk assessment come up. For example, risk assessment came into sharp focus during incident investigations of major accidents in British Petroleum's Texas City Refinery, Buncefield fuel storage and Jaipur tank farm, India. Oil and gas industry typically use Quantitative Risk Assessment (QRA) methodology to analyze and understand risks in its facilities. The method started in nuclear industry and was later adopted in process as well as oil and gas industries. Based on practice of more than 25 years, the industry is aware of the limitations of the method.

Bayesian Network (BN) or Bayesian Belief Network (BBN) is being applied productively as probabilistic risk assessment method in several areas like medicine, computer science, ecology and chemical industry. The method offers certain advantages over QRAs and reveals a fuller risk profile with causes and effects. This book focuses on the application of BN methods to assess risk of major hazards in oil and gas industry. Specifically, the book presents BN models for the major hazards for oil and gas separator, atmospheric hydrocarbon storage tanks, hydrocarbon pipelines and centrifugal compressors and pumps. Examples of the BN models' simulation with generic data are also given.

1.1 Application of BNs for Risk Assessment

BN have been applied with good results in the areas such as computer science, ecology, medicine and chemical industry (Fenton and Neil, 2012; Pouret, Naim, and Marcot, 2008; Kjaerulff and Madson, 2008; Korb and Nicholson, 2010; Hubbard, 2009). However, applications of BN to oil and gas facilities have been very limited specifically with regard to the risks due to major hazards in oil and gas facilities. Loss Of Containment (LOC) scenarios constitute major hazards in the oil and gas facilities. Therefore, causal mechanisms and BN have been developed for such scenarios for the more common equipment, namely oil and gas separator, hydrocarbon pipelines, atmospheric hydrocarbon storage tanks and centrifugal compressor and pumps. The BNs are then simulated and analyzed with generic data.

1.2 The Readership

The book is primarily meant for practicing engineers and researchers in the oil and gas industry as well for graduate students in Process Safety who want to explore the risk assessment area. Rather than describing the theoretical aspects of BN in detail, they cover only the extent necessary for the continuity of the topic. Adequate references have been provided for the interested reader to pursue the relevant topics. The practical aspects are emphasized and given in more detail since that will be what the practicing engineers would want to know.

1.3 Major Limitations of QRA

The main motivation for this book arose out of the author's own experience in the oil and gas industry for over three decades. QRA is well embedded in the process safety studies as a tool for land use planning and spacing of critical units in oil and gas facilities CCPS (2001) and Tweeddale, (2003). However, the majority of the QRAs done during the design stage of a facility usually end up in the records section or library shelves. During operational phase there is very little or no attempts to update these QRAs. When changes are made to the facilities, most of the time QRAs are done only for that portion that undergoes change, which has proven to be fundamentally wrong.

Industry and academia are aware of the limitations of QRA. Please see the ASSURANCE project Lauridsen, Kozine, and Markert (2002) and also Pasman and Rogers (2013); Pasman, Jung, Prem, Rogers, and Yang (2009). The author's personal experience in conducting QRAs also provided the first-hand proof of these limitations. In summary they are as follows:

i. Uncertainties in data for failure frequencies, lack of precision in models and difficulties in identifying common cause failures.

ii. Assumptions are not visible to all concerned.

iii. Models are static, difficulties in capturing variations/changes to the facility

iv. Requires considerable specialist efforts and time

v. Software is costly, calculations are not transparent and limit flexibility

vi. The causes of loss of containment are not investigated in detail

Further comparisons between QRA and BN methodologies are given in Chapter 9.

1.4 BN and Its Advantages

BN is seen as a viable alternative to and/or complementary QRA methodology (Weber, Oliva, Simon, and Iung, 2012; Roy, Srivastava, and Sinha, 2014). As noted earlier, BN is being widely applied to computer science, ecology, finance and chemical industries. Bayesian approaches have also been the subject of a popular book (McGrayne, 2011).

However, oil and gas applications are generally limited to (Kalantarnia, Khan, and Hawboldt, 2010; Kujath, Amyotte, and Khan, 2010; Khakzad, Khan, and Amyotte, 2011; Pasman, and Rogers, 2012; Rathnayaka, Khan, Amyotte, 2012; Khakzad, 2012; Khakzad, Khan, and Amyotte, 2013a, 2013b, 2013c; Cai, Liu, Zhang, Fan, Liu, and Tian, 2013; Abimbola, Khan, and Khakzad, 2014; Ale, Gulijk, Hanea, Hudson, Lin, and Sillem, 2014; Tan, Chen, Zhang, Fu, and Li, 2014).

Main advantages of BN are as follows:

i. It presents the risk in a visually and easily understandable manner

ii. The methodology is transparent.

iii. Failure data and thereby the risk profile can be easily updated in line with changes/updates of the facility

iv. Site-specific data (even if it is sparse) and experts' opinion can be incorporated.

v. Layers of interconnecting causes of loss of containment can be fully explored.

vi. BN can be simulated in predictive and diagnostic mode since causes and effects with their relationships are represented in a transparent model.

In summary, while QRA has its place in Land Use Planning (LUP) and safe distances, Bayesian techniques offer models that can represent a version of cause and effect.

1.5 Scope of the Book

The scope covered in the book is given below:

i. Identification of major hazards and Layers of Protection provided to equipment units in a typical Oil & Gas facility. These are obtained by review of several designs, Piping & Instrument diagrams, Hazard and Operability Study (HAZOP) study & Layers Of Protection Analysis (LOPA) reports from industry.

ii. Development of causal relationship networks for critical equipment/systems failures & its causes, hazards & consequences using the above data

iii. Conversion of these causal relationships to BN.

iv. Simulation of the networks using a suitable software. Testing of the networks with data.

The sources and information required for building the BN is given in Figure 1.1:

In order to develop cause and effect relationships, relevant process safety documents, namely HAZOP, LOPA and Safety Integrity Level (SIL) study reports were studied in detail. These are actual reports from the industry and due to confidentiality are not listed here. Several accident investigation reports were also studied in detail to analyze the root causes that led to such accidents (Lal Committee, 2011; Buncefield Major Incident Investigation

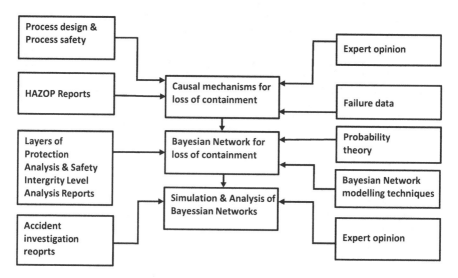

FIGURE 1.1
Sources and information required for building the BN.

Board, 2007, 2008; Herbert, 2010). In parallel, the techniques of developing BN were reviewed to select the right approach to model the cause and effects/influence diagrams (Fenton and Neil, 2012; Kjaerulff and Madson, 2008; Korb and Nicholson, 2010).

BN requires parameterization with failure/incident data. Failure data from several data sources were analyzed to parameterize the BN and the same are given in Table 3.4. In certain cases, expert opinions were also sought. An interdisciplinary approach was required, and materials from many sources (Gulvanessian, and Holický, 2001; Lannoy and Cojazzi, 2002; Bayraktarli, Ulfkjaer, Yazgan, and Faber, 2005; Straub, 2005; Twardy, Nicholson and Korb, 2005; ANU Enterprise-The Murray Darling Basin Authority, 2016) were used to understand the application of BN to risk assessments.

1.6 Structure of the Book

The book is written with nine Chapters. The contents of each chapter are summarized below:

Chapter 1 Introduction provides an overview of the topic as well as the background and motivation for writing this book. Purpose, objectives and scope are given in this chapter.

Chapter 2 Bayes Theorem, Causality and Building Blocks for BN contains basics of probability, description of Bayes theorem and how it can be applied to model cause and effect. A summary of use of discrete and continuous distributions in representing the nature of causes is provided. Further, it describes how complex cause and effect mechanisms can be visually represented as BN using simple building blocks and directed graphs and presents two examples to illustrate the flexibility and power of the BN.

Chapter 3 Bayesian Network for Loss of Containment from Oil and Gas Separator presents the immediate and root (parent) causes for an LOC scenario in a typical oil and gas separator. Causes for LOC as well as the post event scenario are modeled in BN. Application of BN to SIL calculations is also given. Sensitivity feature of BN and how it can be used to find out the sensitivity of other nodes to a target node are illustrated in this chapter.

Chapter 4 Bayesian Network for Loss of Containment from Hydrocarbon Pipeline gives the application of BN to an LOC scenario from a hydrocarbon pipeline. The immediate and root causes as well as the post LOC event scenarios are modeled as BN. Predictive and diagnostic modes of simulating the BN are described. Sensitivities of parent nodes to target node LOC are given. Further, the chapter contains a case study of a natural gas pipeline accident that happened in Andhra Pradesh, India.

Chapter 5 Bayesian Network for Loss of Containment from Hydrocarbon Storage Tank describes the causes and sub-causes for the key causal factors involved in LOC from Floating and Cone roof tanks. Intermediate and immediate causes downstream of the key causal factors as well as its inter-relationships are also defined in the BNs. Post LOC scenarios are modeled for Floating and Cone roof tanks in BN separately. Description of predictive and diagnostics modes of simulation as well as sensitivities of target nodes to other nodes are given here.

Chapter 6 The Jaipur Tank Farm Accident: Retrospective Application of BN to Hazard Assessment of a Floating Roof Tank illustrates the application of BN to model the pre-accident situation existing at a fuel storage terminal to illustrate the predictive nature of the Bayesian approach.

Chapter 7 Bayesian Network for Major Hazards for Centrifugal Compressor Damage presents the immediate and root causes for compressor damage. These have been converted to BN model and predictive and diagnostic modes of analysis as illustrated here. Sensitivities of nodes to target nodes enable fast assessment of the likely contributors to compressor damage.

Chapter 8 Bayesian Network for Loss of Containment from Centrifugal Pump presents the immediate and root causes for LOC from pump. These have been converted to BN model and predictive and diagnostic modes of analysis are illustrated. Sensitivities of nodes to target nodes are given, which enable fast assessment of the likely contributors to pump failure.

Chapter 9 Other Related Topics provides a brief introduction to the techniques of Bayesian inference, which are useful for analysis of failure data and prediction. The chapter also provides a comparison between QRA and Bayesian methodologies.

References lists the books, journal papers, web resources and other material referred in alphabetical order.

2

Bayes Theorem, Causality and Building Blocks for Bayesian Networks

This chapter covers the basics of probability, Bayes Theorem and the nature of causality. It also introduces the application of the Bayes Theorem to cause and effect and thereby the understanding of incidents and accidents. Building blocks of Bayesian Networks (BN) are then introduced. Subsequent sections describe certain cases in a simplistic manner, to enable the reader to quickly understand the power and flexibility of BNs.

2.1 Probability Basics

There are three main approaches to probability, namely classical, frequentist and Bayesian. They are summarized below with an example of rolling of a single die having 6 faces and numbered 1, 2, 3, 4, 5 and 6 in each face. It is to be noted that, when we roll a die, though we cannot predict the outcome, we know the set of all the possible outcomes. We define the sample space as the set of all possible outcomes of an experiment such as the above.

2.1.1 Classical approach states that when one rolls a die there is an equally likely chance (six possible outcomes) of any of the numbers showing up.

2.1.2 When one throws a die repeatedly say 1000 times (theoretically infinite times) recording the numbers that come up for each throw, it is possible to find a pattern that each number shows up approximately 1/6 of the total number of throws, that is $1000 \times 1/6 = 167$ times. If the experiment is repeated infinite number of times, the number will converge to 1/6. We can then state that the probability of a number, say 4 coming up is 1/6. This is the frequentist approach.

2.1.3 Bayesian approach is subjective and states that it is the experimenter's belief that the probability of a number coming up is 1/6.

The first two are known as objective methods, while the Bayesian approach is known as a subjective method.

The basic axioms of probability are given here without further explanation for reference and continuity.

Axiom 1. When we consider a sample space S having n number of events, the probability of any event is at least 0. That is for every event A, the P (A) ≥ 0. There is no such thing called negative probability.

Axiom 2. The probability of the entire sample space is 1 or 100%, since the sample space S contains all the possible outcomes.

Axiom 3. If A_1 and A_2 are mutually exclusive or disjoint events, then the probability of either of them happening is given by $P(A) = P(A_1 \cup A_2) = P(A_1) + P(A_2)$

Based on the above axioms, certain theorems on probability can be written. They are given below without proof.

Theorem 1. The probability of the complement of an event is equal to one minus the probability of the event. That is, if complement is written as ' ¬ A', then $P(\neg A) = 1 - P(A)$. ¬ A is sometimes written as A'.

Theorem 2. For any two events in the sample space, the probability of either event that is either A or B, happening is the sum of the probabilities of the two events minus the probability of both the events happening together. That is $P(A_1 \cup A_2) = P(A_1) + P(A_2) - P(A_1 \cap A_2)$.

Readers may note the difference between the definitions in Axiom 3 and Theorem 2 is the word 'mutually exclusive'. In Theorem 2 the Right Hand Side (RHS) is the probability of either of the event happening and does not consider A and B happening together.

The above two theorems are easily visualized in Figure 2.1.

Independent and Conditional events:

An event A is independent when the probability of occurrence of the event P(A) is not influenced by another event B whose probability is P(B). Suppose we conduct two experiments, flipping a coin and rolling a die, one after the other. These are two separate events.

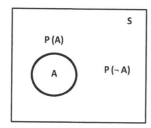 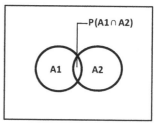

FIGURE 2.1
Theorem 1 and 2 of probability.

Probability of getting a head when flipping of a coin: P(H) = 1/2

Probability of getting a 3 when rolling a die: P(3) = 1/6

Probability of getting a head when flipping of a coin does not affect the probability of getting a 3 when rolling a die. It can be seen that these two events are independent. However, when we want to know the probability of two events happening jointly, that is:

Probability of getting a head P(H) in coin flipping and probability of getting a 3 when rolling a die, P(3), then it will be multiplication of the probability of the first event with the probability of the second event.

That is $P(A \text{ and } B) = P(A) \times P(B)$

Which is the multiplication rule of probability.

Now what happens when a first event can influence a second event?

For example, doctors conclude that if the patient has the presence of a set of symptoms, then she has a particular disease. That means that the probability that the patient has a certain disease depends on the presence of a set of symptoms.

Let us consider the probability of an event A happening given that event B has happened. This can be visualized using the Venn diagram given in Figure 2.2.

Let us consider the probability of an event B, P(B), and probability of an event A, P(A), happening in a sample space S. To compute the probability of an element of A happening once we know that B has happened, we treat the B as a new sample space, and look only at the elements of the sample space that belong to event B. Then we look at the elements in A that are also in B. These elements are in their intersection (that is probability of A and B happening together) and can be written as

$$P(A \text{ and } B) = P(A \cap B) \tag{2.1}$$

The probability of getting an outcome in the intersection event A ∩ B is given by

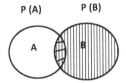

P (A)   P (B)

Inside the intersection P(A)
depends on P(B), that is
P(A) ∩ P(B)

FIGURE 2.2
Venn diagram representation of event A, given B has happened.

$$\frac{\text{Number of outcomes in } A \cap B}{\text{Number of outcomes in } B} = \frac{\dfrac{\text{Number of outcomes in } A \cap B}{\text{Number of outcomes in } S}}{\dfrac{\text{Number of outcomes in } B}{\text{Number of outcomes in } S}}$$

$$= \frac{P(A \cap B)}{P(B)} \tag{2.2}$$

We write the conditional probability by the symbol '|'. For example, the above situation can be written as $P(A|B)$ stated as probability of A given B.

$$P(A \mid B) = \frac{P(A \cap B)}{P(B)} \tag{2.3}$$

The conditional probability $P(A|B)$ is called posterior probability because it gives the probability of the event A after the event B has already occurred. The probability of $P(A)$ is known as a prior probability because it gives the probability of the event A before anything else has happened.

Rearranging the above:

$$P(A \cap B) = P(A \mid B) \times P(B)$$

Extending the principle to three variables:

$$P(A \cap B \cap C) = P(A \mid BC) \times P(BC) = P(A \mid BC) \times P(B \mid C)P(C)$$

which can be generalized as

$$P(A_1, A_2 \ldots A_n \mid) = P(A_1 \mid A_2, A_3 \ldots A_n)\, P(A_2 \mid A_3, A_4 \ldots A_n)$$

$$P(A_{n-1} \mid A_n)\, P(A_n) \tag{2.4}$$

This is called the Chain Rule and provides a means of calculating the full joint probability distribution.

2.1.1 Law of Total Probability

The sample space S can consist of several partitions. For example, when we have a forest in a country, which is the total sample space, that is spread in three distinct districts, the situation can be represented by the Venn diagram as a union of several mutually disjoint events, as shown in Figure 2.3. Here the sample space S is divided in to three districts S1, S2 & S3. The forest sample space is A and it is spread across S1, S2 & S3.

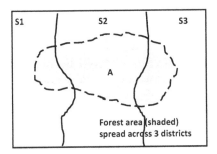

FIGURE 2.3
Forest area A is spread across three districts S1, S2 and S3.

Then the situation can be stated as
Total area of country:

$$S = S1 \cup S2 \cup S3$$

Total area of forest:

$$A = AS = A \cap (S1 \cup S2 \cup S3)$$

$$= (A \cap S1) \cup (A \cap S2) \cup (A \cap S3)$$

These events are mutually disjoint and form the partition of A. Therefore, in terms of probability, the probability of A can be written as

$$= P(A) = P(AS) = P(AS1) + P(A\,S2\,) + P(AS3) \tag{2.5}$$

Now using the Multiplication Theorem, conditional probability P (AS1) can be written as

$$P(AS1) = P(A \mid S1)\,P(S1)$$

Generalizing for n sample spaces:

$$P(A) = \sum_{i=1}^{n} P(A \mid Si)\,P(Si) \tag{2.6}$$

It says that we can calculate the probability of an event if we know the conditional probabilities of each of the events of the partition.

2.1.2 Bayes Formula for Conditional Probability

Bayes formula is given below without rigorous proof:

$$P(A \mid B) = \frac{P(B \mid A)\,P(A)}{P(B)} \tag{2.7}$$

The term P(A|B) is called the posterior probability which is computed by knowing the prior probabilities of A & B denoted by P(A) & P(B) and $P(B|A)$ called likelihood function for B given A.

Bayes theorem is the basis for probabilistic reasoning. That is, given a prior model of the world in terms P(A) & P(B), and a new evidence in terms of likelihood $P(B|A)$, it can decrease my ignorance about the world by $P(A|B)$. The denominator on the RHS can be calculated as total probability of P(B). Please note that this includes false positives and true positives.

$$P(B) = \frac{P(B)P(A)}{P(A)} + \frac{P(B)P(A')}{P(A')} \tag{2.8}$$

where A′ stands for 'Not A'.

Bayes formula also gives a method to test a hypothesis using conditional probabilities. A hypothesis is a suggested explanation for a specific outcome.

We start with a hypothesis H_i, for which our belief is $P(H_i)$ called prior belief about H_i.

Using evidence E, about H_i we can revise and recalculate our belief of H_i as $P(H_i|E)$-called posterior belief about H_i. Using Bayes theorem this will be

$$P(H_i|E) = \frac{P(E|H_i)\, P(H_i)}{P(E)} \tag{2.9}$$

Computing RHS is possible since the likelihood function $P(E|H_i)$, called likelihood of evidence, is often known.

For example, we hypothesize that if a person has toothache then she has good chance of having a cavity (diagnostic probability):

$$P(\text{cavity} \mid \text{toothache})$$

We can try to compute this from $P(\text{toothache}|\text{cavity})$, which is causal probability and prior probabilities of P (cavity) & P (toothache).

Another example is that, when we see a probability P(A|B) high, we can hypothesize that the event B is a cause for event A. Bayes formula can be used to find out the probability of P(A|B), when we know conditional probabilities of the form P(B|A) and prior probabilities of P(A) and P(B).

This way Bayes theorem can be used to describe cause and effect or causality. A typical example is a case in a process facility where there are many spurious trips. One of the causes suggested is fault of the high-pressure sensor in a particular vessel, which can be written as

$$P(\text{Trip} \mid \text{Fault of sensor}).$$

Here in this case we have a sample space that is divided into two, the event that the sensor is faulty and the event when the sensor is not faulty. Now we have the specific case of a trip and we want to hypothesize which event in the partition the specific outcome came from.

We will examine the nature causality further, in terms of cause and effect, and represent the same by Bayes theorem and then discuss BN.

2.2 Bayes Theorem and Nature of Causality

Bayes theorem states that if the probability of occurrence of A and B are stated as P(A) and P(B), then P(A) happening given that B has already happened can be written as

$$P(A \mid B) = \frac{P(B \mid A) P(A)}{P(B)} \tag{2.10}$$

Equation 2.10 can be rewritten as in Equation 2.11 for cause and effect, given that normally we see only the effect.

$$P(\text{effect} \mid \text{cause}) = \frac{P(\text{cause} \mid \text{effect}) \, P(\text{effect})}{P(\text{cause})} \tag{2.11}$$

It states that the probability of an effect, given that a cause has happened, can be described by a combination of the probability of cause given the effect – which would be observable – and the unconditional probabilities of effect and cause. In the RHS of Equation 2.11, P(effect) is the prior probability, P(cause | effect) is the likelihood and P(cause) is the total probability of cause. RHS when computed will give the Left-Hand Side (LHS) known as posterior probability. In the RHS, the denominator of the Equation 2.12 requires calculation of the total probability of cause.

$$P(\text{cause}) = \frac{P(\text{effect}) \, P(\text{cause})}{P(\text{effect})} + \frac{P(\text{effect}) \, P(\text{no cause})}{P(\text{no effect})} \tag{2.12}$$

The relationship is shown schematically in Figure 2.4.

Major hazard in an oil and gas facility is the Loss Of Containment (LOC). Once LOC happens, it could lead to jet fire, vapor cloud explosion or flash fire, pool fire and/or toxic cloud dispersion. With Equation 2.12, relationships can be built up for all the identified causes and effects for LOC of the selected equipment. Different types of relationships are shown in Figure 2.5.

In serial connections as in Figure 3.2b, hard evidence entered at C2 is transmitted to E, at the same time blocking any evidence from C1 reaching E.

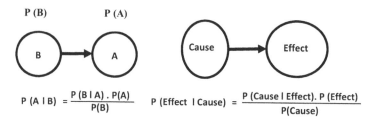

$$P (A \mid B) = \frac{P (B \mid A) . P(A)}{P(B)} \qquad P (Effect \mid Cause) = \frac{P (Cause \mid Effect). P (Effect)}{P(Cause)}$$

FIGURE 2.4
Bayes theorem for cause and effect.

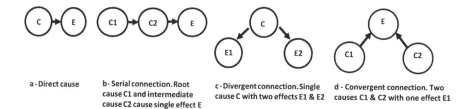

| a - Direct cause | b - Serial connection. Root cause C1 and intermediate cause C2 cause single effect E | c - Divergent connection. Single cause C with two effects E1 & E2 | d - Convergent connection. Two causes C1 & C2 with one effect E1 |

FIGURE 2.5
Types of cause and effects relationships and their Bayesian representation.

This is called d-separation. In other words, C1 & E are d-separated given C2 [1], and this aspect plays an important role in computing BN.

Causes and effects are typically modeled with influence diagrams/fault trees and event trees. The Bow tie diagram is a combination and represents fault tree on LHS and event tree on the RHS. The LOC event is at the center.

2.3 Bayesian Network (BN)

A BN is a directed acyclic graph (DAG) in which the nodes represent the system variables and the arcs symbolize the dependencies or the cause–effect relationships among the variables. A BN is defined by a set of nodes and a set of directed arcs. Probabilities are associated with each state of the node. The probability is defined a priori for a root (parent) node and computed in the BN by inference for the others (child nodes). Each child node has an associated probability table called conditional probability table (CPT).

Effectively BN is an explicit description of the direct dependencies between a set of variables (Fenton and Neil, 2012).

2.3.1 General Expression for Full Joint Probability Distribution of a BN

When we have a BN of n variables A_1, A_2 ...A_n, using the chain rule the joint probability distribution can be written as

$$P(A_1, A_2...A_n|) = P(A_1|A_2, A_3...A_n) \, P(A_2|A_3, A_4...A_n)$$

$$P(A_{n-1}|A_n) \, P(A_n) \qquad (2.13)$$

which can be written using the product symbol

$$P(A_1, A_2...A_n|) = \prod_{i=1}^{n} P(A_i|A_{i+1}....A_n) \qquad (2.14)$$

However if we know that A_1 has exactly two parents A_3 and A_5, then the generic part of the joint probability of equation's LHS

$$P(A_1|A_2, A_3...A_n)$$

reduces to

$$P(A_1|A_3, A_5)$$

Therefore, in general, if Parents (A_i) denote the set of parents of the node A_i, then the full joint probability distribution can be simplified as

$$P(A_1, A_2...A_n|) = \prod_{i=1}^{n} P(A_i| \text{ Parents } (A_i) \qquad (2.15)$$

2.3.2 Illustrative Example of Application

Application of the above principles will be illustrated in the following two simple BNs for process systems.

A. Emergency Shut Down Valve (ESDV) operation
   An ESDV acts to prevent a hazardous situation from developing into an accident. The situation can be represented as an Event Tree given in Figure 2.6. Let us assume that the probability of an ESDV working is 0.85. The probability values are hypothetical and not from any database. Conversely, the probability of ESDV not working is 0.15. If ESDV works the probability of Safe Shutdown is 0.97. If ESDV does not work the probability of Safe Shut down is only 0.02.

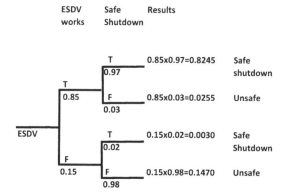

FIGURE 2.6
Event Tree for ESDV action and safe shutdown.

From the Event Tree the following can be calculated:

Probability of Safe Shutdown = 0.8245 + 0.0030 = 0.8275
Probability of Unsafe situation= 0.0255 + 0.1470 = 0.1725

The Even Tree can be converted to an influence diagram shown in Figure 2.7.
    For complex systems a systematic approach will be required. Please see (Lannoy and Cojazzi, 2002) for details.
    The equivalent BN is given in Figure 2.8.
    The BN shows the end results of the Event Tree calculation when input to the conditional probability statements for node 'Safe Shutdown' are given as in Table 2.1 as CPT:
    The BN model in Figure 2.8 shows the forward probabilities which are same as the results from Event Tree. Now we have a situation where we

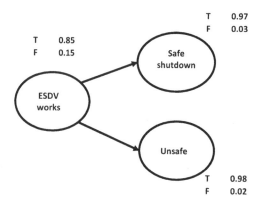

FIGURE 2.7
Influence diagram for ESDV action and safe shutdown.

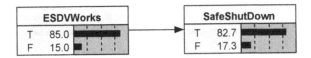

FIGURE 2.8
BN for ESDV action and safe shutdown.

TABLE 2.1

CPT for 'SafeShutdown'

ESDV works	Safe Shutdown	
	T	F
T (0.85)	0.97	0.03
F (0.15)	0.02	0.98

know that Safe Shutdown has occurred. What is the probability that ESDV has worked?

In order to calculate the same, Bayes theorem has to be used which is illustrated below:

Probabilities of Safe Shutdown and No Safe Shutdown, given that ESDV has worked

$$P\frac{\text{Safe Shutdown}}{\text{ESDV works} - \text{True}} = 0.97 \tag{2.16}$$

$$P\frac{\text{Safe Shutdown}}{\text{ESDV works} - \text{False}} = 0.02 \tag{2.17}$$

Applying Bayes theorem for finding the probability ESDV working, given there is Safe Shutdown:

$$P\frac{\text{ESDV Works} - \text{True}}{\text{Safe Shutdown}} = P\frac{\text{Safe Shutdown}}{\text{ESDV works} - \text{True}} \times P\frac{(\text{ESDV Works} - \text{True})}{P(\text{Safe Shutdown})} \tag{2.18}$$

In the above expression, RHS numerator values are known. The unconditional probability of Safe Shutdown P (Safe Shutdown) in the denominator needs to be calculated.

$$P(\text{Safe Shutdown}) =$$

$$P(\text{ESDV Works} - \text{True}) \times P\frac{\text{Safe Shutdown}}{\text{ESDV works} - \text{True}} +$$

$$P(\text{ESDV Works} - \text{False}) \times P\frac{\text{Safe Shutdown}}{\text{ESDV works} - \text{False}} \tag{2.19}$$

$$= 0.85 \times 0.97 + 0.15 \times 0.02 = 0.8275$$

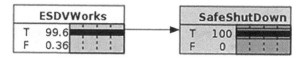

FIGURE 2.9

BN for ESDV action when safe shown is confirmed.

TABLE 2.2

List of Popular Software for BN

SI. No	Name of Software	Company/Organization	Internet Site
1	Analytica	Lumina Decision System Inc.	www.lumina.com
2	Bayesia	Bayesialab	www.bayesia.com
3	GeNIe	Decision System laboratory, University of Pittsberg	http://dslpitt.org/genie/
4	Netica	Norsys Corporation	www.norsys.com
5	Hugin	Hugin Expert	www.hugin.com
6	JavaBayes	University of Sao Paulo	https://www.cs.cmu. edu/~javabayes/index.html
7	MSBNx	Microsoft	http://research.microsoft. com/adapt/ MSBNx
8	AgenaRisk	Agena Ltd.	www.AgenaRisk.com

Substituting the above value in Equation 2.18

$$P\frac{\text{Safe Shutdown}}{\text{ESDV works} - \text{False}} = \frac{0.97 \times 0.85}{0.8275} = 0.9963$$

The above computation can be readily achieved in the Bayesian simulation by changing the Safe Shutdown True to 100%. The computation is propagated backwards using the Bayes theorem to give the result as 0.9963, as shown in Figure 2.9.

Simple situations like the above can be done manually or with a spreadsheet. But complex and large BN require software. Netica (Netica, 2017) is used in simulations shown in this book. Several other software are available for BN simulation. List of most popular software for BN is given in Table 2.2.

2.4 Oil and Gas Separator

For the reader to easily understand the methodology BN followed in this book, a simplified version of LOC from a pressure vessel, namely that of an oil and gas separator, is given in this section. A detailed description and

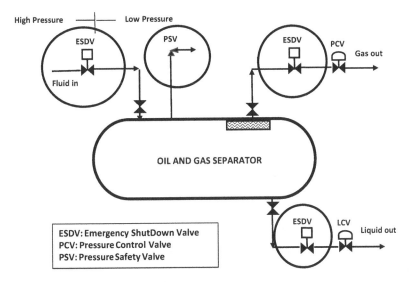

FIGURE 2.10
Typical oil and gas production separator.

analysis are available in Chapter 3. Figure 2.10 shows a typical upstream oil and gas production separator with critical safety barriers. The inlet is a reservoir fluid consisting of oil, gas and water. The separator is envisaged as a device to separate gas and liquid, with outlets for each of them. Figure 2.11 shows an example of a Bow tie diagram for a simplistic and illustrative Fault Tree and Event Tree of an LOC as a separator.

In Figure 2.11, the Fault Tree on the left-hand side depicts the potential hazards and its corresponding mitigating measures and is built up using OR and AND gates. The probability of LOC can be calculated when probabilities are assigned to each of the failures. From LOC, the event can proceed to any of the scenarios in the event tree. With probabilities being known for each branch in the event tree, the probability of each end consequence is calculated.

Using the connections described in Figure 2.11, the equivalent BN to the above Bow tie has been developed and the same is shown in Figure 2.12. Mapping of the Fault Tree, Bow tie and Event tree is described in literature (Bobbio, Portinale, Minichino, and Ciancamerla, 2001; Khakzad, Khan, and Amyotte, 2013a) and in Unnikrishnan, Shrihari, and Siddiqui (2014). The compact model given in the above paper is used here. The software used is Netica®.

Once the BN has been finalized, it is parameterized with the probability of failure values in the parent nodes (nodes without any predecessors). The probability values used in the illustrative BN is given in Table 2.3.

The conditional relationships, OR and AND gates in this case, are encoded in the CPTs of the child nodes. CPTs for the child nodes OverPressure and FailureOfSepLOC are given in Table 2.4.

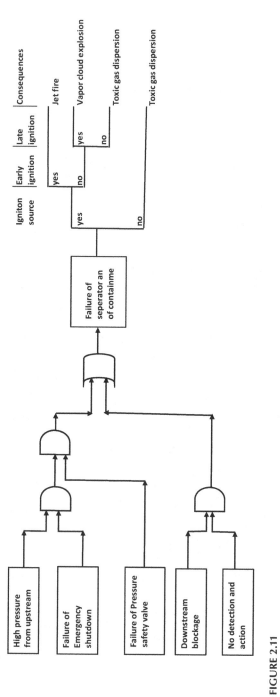

FIGURE 2.11
Separator LOC and Bow tie.

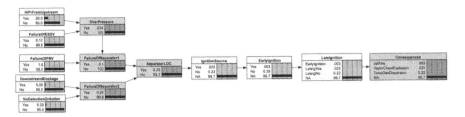

FIGURE 2.12
BN network for Bow tie diagram of separator.

TABLE 2.3

Details of Parent Nodes for Separator LOC

SI. No.	Node Name	Node Full Form	States and Probability Value	Parameterization Method	Description
1	HiPrFromUpstream	High pressure from upstream	Yes [0.20] No [0.80]	Discrete (Manual)	High pressure can come from upstream well side
2	Failure of ESDV	Failure of ESDV	Yes [0.00165] No [0.998]	Discrete (Manual)	The probability of failure on demand for ESDV is used.
3	Failure of PSV	Failure of pressure safety valve	Yes [0.001] No [0.0099]	Discrete (Manual)	The probability of failure on demand for PSV is used.
4	DownstreamBlockage	Downstream blockage	Yes [0.005] No [0.0095]	Discrete (Manual)	There could be downstream blockage from demister or valves
5	NoDEtectionOrAction	No detection or action	Yes [0.005] No [0.0095]	Discrete (Manual)	Operator may not detect the downstream blockage & take action

The BN basically describes the joint probabilities of the events and can be used for several types of analysis. Predictive mode will calculate the probability of occurrence of the event, namely LOC. On the other hand, if LOC can be assumed to have occurred, by making the 'Yes' state in the node FailureOfSepLOC 100%, then the state values for corresponding parents are backcalculated by the software. This is the diagnostic mode, which is useful for understanding the causes that would have contributed to a larger

TABLE 2.4

CPT Example

Over Pressure	High PrFrmUpstream	Failure OfSDV	Failure Of5epLOC	Failure OfSeparator1	FailureOf Separator2
Yes	Yes	Yes	Yes	Yes	Yes
No	Yes	No	Yes	Yes	No
No	No	Yes	Yes	No	Yes
No	No	No	No	No	No

a-CPT for child node OverPressure (AND gate)

b-CPT for child node Failure Of SepLOC (OR gate)

FIGURE 2.13
BN with LOC 100%.

extent in causing the event. The corresponding result of the separator model in Figure 2.13, with an LOC equals 100%.

The model with a new evidence of LOC equals 100%, which indicates that the main casual factor could be downstream blockage with no detection or action. The flexibility and power of BN are evident from the above example.

Using the principles described above, BN for oil and gas separator, hydrocarbon atmospheric storage tank, hydrocarbon pipeline and compressor have been developed. CPTs for the BNs have been populated with failure data taken from various sources that are listed in Table 4.1.

2.5 Sensitivity to Findings

It would be of interest to know how the changes in values of child nodes (findings of effect nodes) can affect the parent nodes. One way of doing this is to manually change the value of the probability (findings) of child nodes and to see the how the probabilities change at the parent nodes in BN. An easier way will be to use the tool 'Sensitivity to Findings' available in Netica (Netica®, 2017).

In other words, sensitivity analysis is a tool that can be used to study how the variation (or uncertainty) in the output (child nodes) of a model can be apportioned to different sources of variation in the input (parent nodes) of a model. Through sensitivity analysis, variables or parent nodes that have the highest influence in BN models effect (Child) nodes as well as its relative importance, can be obtained.

Two types of sensitivity analyses can be used in evaluating a BN. The first type 'Sensitivity to findings' considers how the BN's posterior distributions change under different conditions, while the second type 'Sensitivity to parameters' considers how the BN's posterior distributions change when parameters are altered. In the Netica® version used in this research only 'Sensitivity to findings' is available, and the same has been used to find out which of the nodes have the highest impact on the LOC. The following description is based on Korb and Nicholson, (2010) and Kjaerulff and Madison, (2008).

Sensitivity to findings uses two types of measures: entropy reduction or mutual information for discrete variables and variance reduction for continuous variables.

Entropy is a measure of randomness. The more random the variable is, the higher its entropy will be. In other words, Entropy is a measure of how much the probability mass is scattered over the states of a variable (the degree of chaos in the distribution of the variable).

If X be a discrete random variable with n states $x_1, x_2, \ldots x_n$ and probability distribution of X is P (X), then the entropy of X is defined as

$$H(X) = -\sum_X P(X) \log P(X) \qquad (2.20)$$
$$\geq 0$$

where log is to the base 2

The mutual information of X and Y is denoted as I (X, Y).

The conditional entropy H (X | Y) is a measure of the uncertainty of X given an observation on Y, while the mutual information I (X, Y) is a measure of the information shared by X and Y (Netica, 2017). If X is the variable of interest, then I (X, Y) is a measure of the value of observing Y. The mutual information is computed as

$$I(X, Y) = H(X) - H(X \mid Y) \qquad (2.21)$$

$$= H(Y) - H(Y \mid X) \qquad (2.22)$$

$$= \sum_Y P(Y) \sum_X P(X \mid Y) \log \frac{P(X, Y)}{P(X)P(Y)} \qquad (2.23)$$

Node	Mutual information
Failure Of Separator 2	0.02517
No Detection Or Action	0.01086
Downstream Blockage	0.01086
Failure Of Separator 1	0.00003
Over Pressure	0
Failure Of ESDV	0
Failure Of PSV	0
Hi Pr From Up stream	0

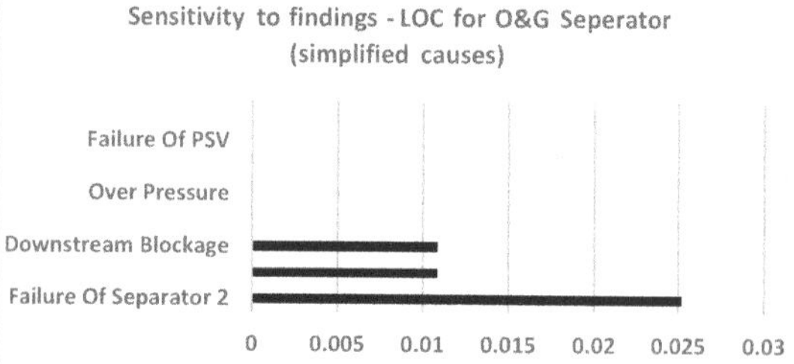

Sensitivity to findings - LOC for O&G Seperator (simplified causes)

FIGURE 2.14
Sensitivity to findings for separator LOC – simplified.

In principle I (X, Y) is a measure of the distance between P (X) P (Y) and P (X, Y). The conditional mutual information given a set of evidence ε is computed by conditioning the probability distributions on available evidence ε:

$$I\left(X, Y \mid \varepsilon\right) = \sum_{Y} P\left(Y \mid \varepsilon\right) \sum_{X} P\left(X \mid Y, \varepsilon\right) \log \frac{P\left(X, Y \mid \varepsilon\right)}{P\left(X \mid \varepsilon\right) P\left(Y \mid \varepsilon\right)} \quad (2.24)$$

$I\left(X, Y \mid \varepsilon\right)$ is computed for each possible observation of Y. Netica® readily calculates the probabilities needed for the computation.

The other measure is the variance, which is a measure of the dispersion of X around mean (Korb and Nicholson, 2010).

$$\text{Var}\left(X\right) = \sum_{x} P\left(x - \mu\right)^2 P\left(x\right) \quad (2.25)$$

where μ is the mean.

The greater the dispersion, the less is known, and the sensitivity between the connected nodes is higher.

Figure 2.14 shows the 'Sensitivity to findings' computed by Netica® for the example of oil and separator for the node 'Separator LOC'. For discrete variables Mutual information is the parameter used as a measure of sensitivity. It can be seen that of all the parent (root) nodes, 'Downstream blockage' and 'No detection or action' have the highest influence on the occurrence of the event. The values have been converted to a bar chart on the right-hand side of the figure.

2.6 Use of Probability Density Functions and Discretization

One of the challenges faced while building BN is the use of Probability Density Functions (pdfs) for discrete and continuous distributions.

In most of the actual cases, failure data of different types of equipment are not available or imprecise. In such situations, it will be prudent to use a pdf that is suited to the equipment. It will be better to understand the equipment problems and maintenance data from an experienced reliability professional before decisions are taken on the basis of BN simulations.

Details of pdfs are available in several excellent references (Montgomery and Runger, 2011; Krishna, 2006), which the readers are encouraged to study. Figure 2.15 gives some of the more commonly used pdfs in BN simulations and their parameters.

Other continuous distributions that can be used are Weibull, Gamma and Beta.

NoisyOr distribution is also used for BN simulations in this book, which is described further.

Continuous distribution needs to discretized for simulation to be done. In Netica® this can be done by manual discretization by defining the thresholds and value ranges. The other option is to use the feature of 'learning from data'. Interested readers can see Ni, Phillips, and Hanna (2011).

Please note the resulting probabilities are sensitive to discretization levels, and therefore the step may have to be repeated to suit the actual data available.

2.7 Framework for BN Application for Major Hazards

The principles of Bayes theorem and network described under earlier sections were applied to the development of BNs for major hazards in oil and separator, atmospheric hydrocarbon storage tanks, hydrocarbon pipelines, centrifugal compressors and centrifugal pumps. Major hazards in these equipment are mainly LOC, except in the case of compressors. Oil and separator, atmospheric hydrocarbon storage tanks and hydrocarbon pipelines have LOC scenarios that can lead to high consequence accidents. LOC of the compressor itself is very rare, and therefore damage is considered as a major hazard for compressor. Leakage of process gas is considered separately in the event tree. Once the BN is defined and constructed using causal relationships, the parent nodes need to be parametrized with probability data. These are obtained from the following sources.

2.8 Sources of Failure Data

2.8.1 Published Data

A comprehensive survey was undertaken to identify the sources for failure data for the equipment under consideration. Table 2.5 lists the main sources that were used in the development of BN.

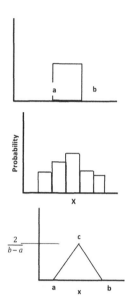

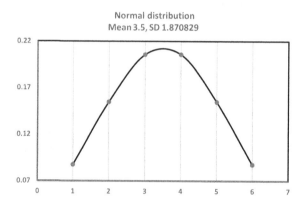

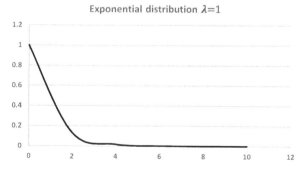

FIGURE 2.15
Some of commonly used pdfs in this book.

TABLE 2.5

Sources for Failure Data

Organization	Reference Documentation and Citation
International Association of Oil & Gas Producers	Process release frequencies Report No. 434-1. (International Association of Oil & Gas Producers, 2010)
International Association of Oil & Gas Producers	Ignition probabilities Report No. 433–6.1 (International Association of Oil & Gas Producers, 2010)
International Association of Oil & Gas Producers	Storage incident frequencies Report No. 434-03 ((International Association of Oil & Gas Producers, 2010)
Offshore Reliability Data	OREDA Handbook (OREDA, 2002)
Det Norske Veritas (DNV): Failure frequency guidance	Det Norske Veritas (DNV). (Det Norske Veritas (DNV), 2013)
Center for Chemical Process Safety	Layers of Protection Analysis (Center for Chemical Process Safety, 2001)
European Gas Pipeline Incident Group (EGIG)	Report of the European Gas Pipeline Incident Data Group, (European Gas Pipeline Incident Data Group, 2014)
USDOT PHMSA database	Review of the US Department of Transportation Report (Stover, 2013)
CONCAWE database	Performance of European cross-country oil pipelines, Report No. 12 (Davis, Diaz, Gambardella and Uhlig 2013).
Marshal & McLennan	Atmospheric storage tank-Risk engineering position paper-01 (Marsh and McLennan, 2011)
Large Atmospheric Storage Tank Fires Group	Large Atmospheric Storage Tank Fires (LASTFIRE, 2001)
Institut National de l'Environnement Industriel et des Risques (INERIS)	Accidental Risk Assessment Methodology for Industries (ARAMIS). (INERIS, 2004)
Flemish Government	Handbook of Failure Frequencies 2009 for drawing up a safety report (Flemish Government, 2009)
Oil India Safety Directorate	Report of the Committee on Jaipur Incident (Lal Committee, 2011)
Buncefield Major Incident Investigation Board	Recommendations on the design and operation of fuel storage tanks. (Buncefield Major Incident Investigation Board, 2007). The Buncefield Incident 11 December 2005-The Final Report (Buncefield Major Incident Investigation Board, 2008)
E&P Forum	Hydrocarbon Leak and Ignition Database Report 11.4/180. (E&P Forum, 1992)
TNO (VROM) –Netherlands.	Guidelines for Quantitative Risk Assessment, Purple Book, CPR 18E (TNO -VROM, 2005)
Health and Safety Executive UK	Failure Rate and Event Data for use within Risk Assessment (Health and Safety Executive, 2012)

2.8.2 Industry Reports

Apart from the above, several industry study reports and documents on Hazard and Operability Studies (HAZOP), Layers Of Protection Analysis (LOPA) and Safety Integrity Level (SIL) were studied to understand the identified failure mechanisms and failure rates used in practice. They are not listed due to the confidential nature of contents.

2.9 Chapter Summary

Basics of Bayes theorem and how it can be applied to causal mechanisms are illustrated in this chapter. Manual calculations are given for predictive forward calculations typical of an Event Tree and for diagnostic (backward) calculations, given an evidence using Bayes theorem. BN and how it can give both these calculations are shown. Two examples of simple BN – ESDV action and oil and gas separator hazards – are described to bring out the methodology that is implemented in the subsequent chapters. The feature of 'Sensitivity to a finding' of the Netica® software is described. This feature allows determination of contribution of each parent node to a finding at a target (child) node. The framework of how the above is employed for developing BNs for LOC of the equipment under consideration is given in the last section.

The ensuing chapters describe the development and evaluation of BN applications to the most common equipment in oil and gas industry.

3

Bayesian Network for Loss of Containment from Oil and Gas Separator

3.1 Oil and Gas Separator Basics

Oil and gas production separator receives the reservoir fluid from the oil well and separates it into oil, gas and water. It can be a three-phase (oil, water and gas) or only a two-phase (liquid and gas only) separation. Oil and gas separators are usually designed for a lower pressure than that of the shut off pressure of the well due to economic reasons, and sufficient layers of protection are provided for the vessel to mitigate the risk of an overpressure. Figure 2.10 is reproduced here as Figure 3.1, for continuity, showing a typical oil and separator with its layers of protection for easy reference.

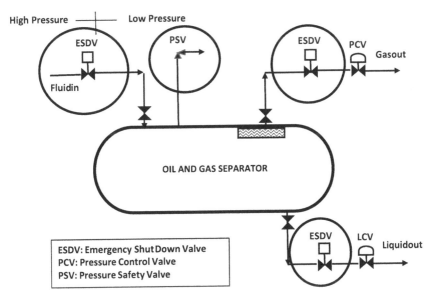

FIGURE 3.1
Oil and gas separator and its layers of protection.

3.2 Causes for Loss of Containment

The main causes for loss of containment (LOC) from a typical oil and gas separator are due to overpressure and leakage of flammable gas or liquid. Ignition of the same can result in serious fire accidents. Fire near the separator due to any extraneous causes could weaken the mechanical integrity of the vessel and in LOC. All causes considered are shown in the Fault tree (FT) diagram in Figure 3.2. The nodes and their states used in BN are given in Table 3.1. The Fault tree has been converted to BN using established techniques (Bearfield, and Marsh, 2005; Duan, and Zhou, 2012). The BN itself is shown in Figure 3.3.

3.3 Bayesian Network for LOC in Oil and Gas Separator

Table 3.1 depicts the causes and its mitigation measures typically employed for an industrial oil and gas production separator. The equivalent BN for LOC is given in Figure 3.3.

The node 'LossOfContainment' is an OR gate that combines the parent nodes as per the equation 3.1:

P (LossOfContainment | Failure1OfSeparator, Failure2OfSeparator,

LossOfSeparatorIntergrity, CatastrophicVesselFailure, IntLOC)

= (Failure1OfSeparator || Failure2OfSeparator || LossOfSeparatorIntergrity ||

CatastrophicVesselFailure || IntLOC)   (3.1)

As can be seen from the BN, given the probabilities of parent and child nodes, on a relative scale, the probability of LOC is 0.00024 (in BN in the figure they are given in percentages).

This relative probability can be considered similar, and scaled to correspond with the industry average value of probability of failure of 5×10^{-6} failures per year given in Toegepast Natuurwetenschappelijk Onderzoek (Netherlands Organization for Applied Scientific Research), TNO Purple book and assuming a constant failure rate (TNO (VROM), 2005). Another source of release data is International Association of Oil & Gas Producers Report No. 434-1, Process release frequencies (International Association of Oil & Gas Producers, 2010).

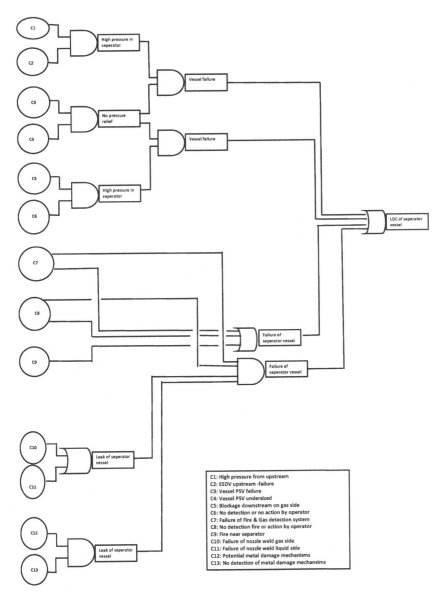

FIGURE 3.2
Fault tree for LOC in oil and gas separator.

On predictive mode of the BN, any change of the values in states of the parent nodes will impact the value of the last child node, LOC.

The Event tree (ET) for LOC can be readily mapped into BN (Pouret, Naim, and Marcot, 2008) [2], (Kjaerulff and Madson, 2008) [3]. The LOC can be in

TABLE 3.1

Details of Parent Nodes for LOC in Oil &Gas Separator

SI No.	Node Name	Node Description	States & Probability Value	Parameterization Method	Description
1	C1	High pressure from upstream	Yes [0.20] No [0.80]	Manual	High pressure can come from upstream well side
2	C2	ESDV –upstream failure	Yes [0.00165] No [0.998]	Manual	The PFD for ESDV is used.
3	C3	Vessel PSV Failure	Yes [0.001] No [0.99]	Manual	The PFD for PSV is used.
4	C4	Vessel PSV undersized	Yes [0.001] No [0.0099]	Manual	PSV could be undersized
5	C5	Blockage downstream on gas side	Yes [0.005] No [0.95]	Manual	There could be downstream blockage from demister or valves
6	C6	No detection or no action by operator	Yes [0.005] No [0.95]	Manual	Operator may not detect the downstream blockage & take action
7	C7	Failure of Fire & Gas detection system	Yes [0.001] No [0.99]	Manual	Fire & Gas (F&G) may have malfunctioned or failed.
8	C8	No detection fire or action by operator	Yes [0.001] No [0.99]	Manual	Operator may not detect fire or gas leak.
9	C9	Fire near separator	Yes [0.008] No [0.92]	Manual	Operator may not detect fire or gas leak.
10	C10	Failure of nozzle weld gas side	Yes [0.05] No [0.95]	Manual	Gas side outlet nozzle weld could leak
11	C11	Failure of nozzle weld liquid side	Yes [0.005] No [0.95]	Manual	Gas side outlet nozzle weld could leak
12	C12	Potential metal damage mechanisms	Yes [0.006] No [0.94]	Manual	Damage mechanisms like sulfide stress cracking (SSC), hydrogen induced cracking (HIC), etc. could be progressing.
13	C13	No detection of metal damage mechanisms	Yes [0.005] No [0.95]	Manual	Inspections may not detect SSC or HIC

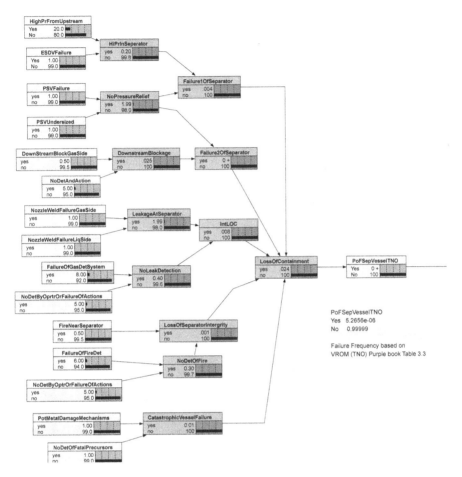

FIGURE 3.3
BN for oil and gas separator LOC FT.

gas phase or in two phases. The series of possible events that can happen in sequence after an LOC is ignition with sub-probabilities early or late in both cases. The ignition probabilities are based on (IOGP, 2010). The end effects in each of the cases are given at the end of the ET as shown in Figure 3.4. The corresponding BN is given in Figure 3.5.

Fault tree and ET can be combined as a BN and is given in Figure 3.6. In this case, in order to get the exact match with the published data of probability of failure, the BN node for LOC has been scaled to another node next to it titled 'PoFSepVesselTNO' with 5×10^{-6} failures per year.

Further, the BN can work in diagnostic mode. If there is an LOC, the yes state in that node is set to 100%. Then the BN recalculates the values of all

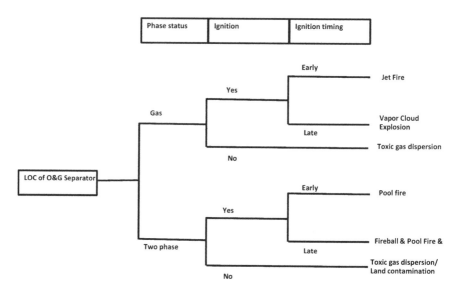

FIGURE 3.4
ET for consequences of LOC in an oil and gas separator.

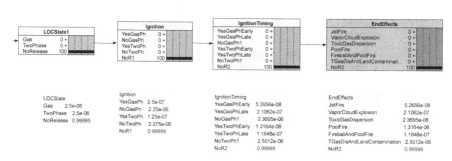

FIGURE 3.5
BN for consequences of LOC in an oil and gas separator.

the nodes. The diagnostic mode is given in Figure 3.7. It states that the most probable cause of an LOC is the failure of no detection of potential damage mechanism followed by leak from gas or liquid side piping.

Various analysis can be done on the BN. For example, with the LOC as 100%, then the probability of a vapor cloud explosion will jump from 0.0000007 to 0.00088.

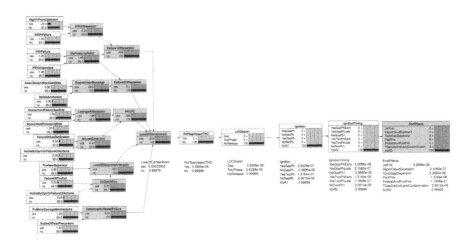

FIGURE 3.6
BN for LOC and its consequences, FT and ET combined.

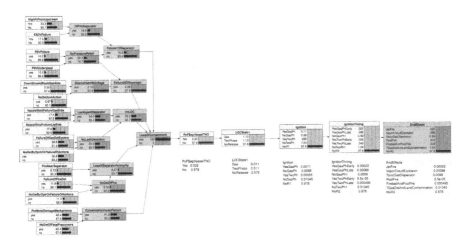

FIGURE 3.7
BN for LOC – Diagnostic mode. LOC is set to 100%.

3.4 Sensitivities

Netica® can calculate parameters for sensitivities of findings at other nodes to a target parameter called query node. A typical graph showing sensitivities of other nodes to LOC node is given in Figure 3.8.

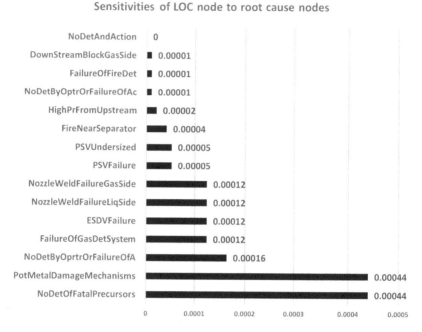

FIGURE 3.8
Sensitivity of LOC node to root cause nodes.

As expected, it shows that, given the current prior probabilities, the highest change in the relative probability of LOC will happen when the potential for damage mechanisms and 'no detection of fatal precursors' coexists.

3.5 Application of BN to Safety Integrity Level Calculations for Oil and Gas Separator

BN can be applied to analyze independent protection layers (IPL) and Safety Instrumented Function for oil and gas separators. American Institute of Chemical Engineers defines IPL as 'A device, system, or action that is capable of preventing a scenario from proceeding to the undesired consequence without being adversely affected by the initiating event (IE) or the action of any other protection layer associated with the scenario'.

Figure 3.9 shows the layers of protection for a three-phase oil and gas separator.

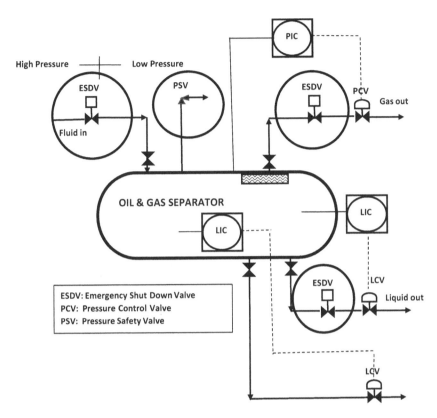

FIGURE 3.9
Three-phase oil and gas separator with the independent layers of protection.

3.5.1 The Independent Protection Layers (IPLs)

i. IPL1: Adequate process and mechanical design of the separator vessel is the first layer of protection. However in typical Safety Integrity Level (SIL) calculation, the Probability of Failure on Demand (PFD) for this layer is taken as 1, PFD=1.0. Node name: ProcessDesignFailure.

ii. IPL2: Basic Process Control System (BPCS) – the Pressure Control Valve (PCV) for controlling the vessel pressure. Node name: BPCSFailure, PFD value is 0.01.

iii. IPL 3: The last layer of protection is the Pressure Safety Valve (PSV) for letting the gas out to the flare in case the pressure goes up beyond the set point. Node name: PSVFailure. PFD value is taken as 0.01.

iv. IPL4: The Safety Instrumented System (SIS) forms the next IPL; namely the Emergency Shutdown Valve (ESDV) comes into action independently once the BPCS and Operator action have failed.

SIL calculations are done to find out the level of reliability that has to be built into the SIS (the full ESDV loop) considering the declared Risk Tolerability policy of the company.

The Pressure Alarm High (PAH) alarm coming from the control system (not shown in BN) is meant to initiate Operator action to control the sudden rise in pressure. However, Operator action is not considered as an IPL in this study. Depending on the company's policies this IPL may be included in SIL calculations.

3.5.2 ET for Layer of Protection Analysis (LOPA)

An IE and the Layers of Protection provided for the separator and its sequential failure can be depicted as an ET. It is shown in Figure 3.10 together with its associated calculations. IPL iv. is not included in the initial calculations.

The above ET can be readily mapped into BN as described in Chapter 2. Figure 3.11 gives the BN for Layers of Protection as well as the required PFD calculations for the Safety Instrumented Function.

The calculations show that the Mitigated Failure Frequency (MFF) is 1e-06, which in this case is equal to the declared Individual Risk Tolerability

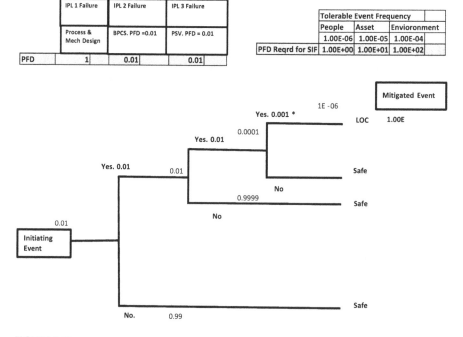

FIGURE 3.10
ET for layers of protection analysis for oil and gas separator.

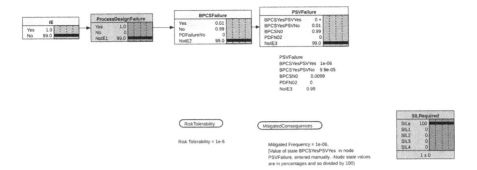

FIGURE 3.11
BN for oil and gas separator SIL calculations.

Criteria (IRTC) for the company (1e-06) considered under this study. Then, the PFD required for the Safety Integrity Function (SIF) can be calculated as

$$PFD \text{ requried for } SIF = \frac{IRTC}{MFF} = \frac{1e\text{-}06}{1e\text{-}06} = 1$$

As per the SIL table shown in Table 3.2, the above PFD translates to SIL a, or no specific SIL assignment for SIF.

The BN shown in Figure 3.12 is an exact representation of the ET in Figure 3.11 with an added feature of an influential factor that can affect the failure frequency of an IPL. Regular testing of the PSV is an important factor that ensures that the failure rate of the PSV is not deteriorating with respect to time.

This aspect is included in the node PSVTestNotProper, which when confirmed 'yes' or 'no' will affect the PFD of the PSV. Figure 3.12 shows the added node for testing.

The SIL calculations in BN indicate requirement of SIL a, which indicates that no specific SIL assignment is required to achieve the IRTC of 1×10^{-6}.

TABLE 3.2

SIL, PFD, Availability & Risk Reduction Factor

Safety Integrity Level (SIL)	Probability of Failure on Demand (PFD)	Availability Required %	Risk Reduction Factor (RRF)=1/PFD
1	10^{-2} to 10^{-1}	90.00–99.00	100–10
2	10^{-3} to 10^{-2}	99.00–99.90	1,000–100
3	10^{-4} to 10^{-3}	99.90–99.99	10,000–1,000
4	10^{-5} to 10^{-4}	>99.99	100,000–10,000

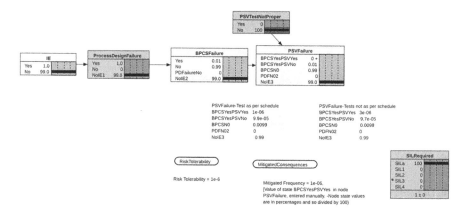

FIGURE 3.12
BN for oil and gas separator SIL calculations with testing.

3.6 Chapter Summary

BN for oil and separator is discussed in this chapter. The root and intermediate causes for an LOC along with mitigation measures are summarized in a table and converted to BN. The usefulness of diagnostic mode of BN simulation is illustrated. Sensitivity of parent nodes to LOC is given, which shows that existence of metal damage mechanisms and 'No detection of fatal precursors' has the highest contribution to an LOC. The method for application of BN to calculate SIL values is described to highlight that all the factors that affect the IPL can be included in the BN.

The development of BN for hydrocarbon pipeline hazards is taken up in the next chapter.

4

Bayesian Network for Loss of Containment from Hydrocarbon Pipeline

4.1 Causes of Pipeline Failures

Pipelines carry crude oil and products such as hydrocarbon liquids and gas from production centers to consumer points. Huge pipeline networks exist in USA, Europe, UK and Canada. In fact, such pipeline networks are critical infrastructures and therefore need to designed, operated and maintained at the highest level of safety. Several agencies have been collecting data on pipelines and causes of pipeline failures, and data have been documented and analyzed by these organizations. Prominent among them are European Gas pipeline Incident data Group (EGIG). (European Gas pipeline Incident data Group, 2014) for gas pipelines, US Department of Transportation Pipeline and Hazardous Material Safety Administration-(US DoT-PHMSA) (Stover, 2013) and CONservation of Clean Air and Water in *Europe*. Association of the *European* downstream oil industry (CONCAWE) for oil pipelines in Europe (Davis, Diaz, Gambardella, and Uhlig, 2013) for liquid pipelines. Their data is available in public domain. Contribution of various causes to the overall pipeline failure is given here from the above references. Figure 4.1a–c depict the distribution of causes of failure for gas and liquid pipeline failures, respectively. Table 4.1 presents the main causes and sub-causes identified in these reports in a tabular form.

The pipeline failure rates itself have been on the decrease throughout the years. In general, the gas pipelines failure rates are in the range of 0.1–0.25/10^3 km year (Marsh and McLennan, 2011). Liquid pipelines failure rates are higher, nearly twice, than that of gas pipelines.

Oil and Gas Processing Equipment

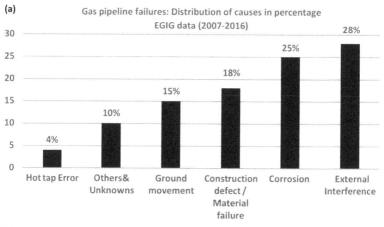

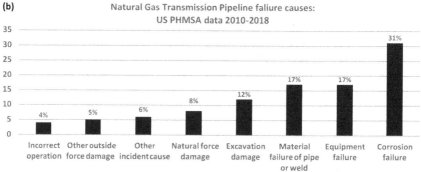

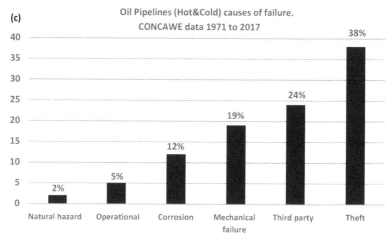

FIGURE 4.1

(a) Distribution of causes for gas pipeline failure EGIG data, (b) Distribution of causes of natural gas pipeline failure PHMSA data and (c) Distribution of causes for liquid pipeline failure CONCAWE data.

TABLE 4.1

Main Causes and Sub-Causes for Pipeline Failures

EGIG (2009–2013)	CONCAWE-Liquid Pipelines (1971–2012)		US Dot PHMSA (2006–2010)
Main Causes	**Main Causes**	**Sub-Causes**	**Main Causes**
1. Construction defects/material failures	1. Mechanical failure	• Design • Construction • Materials fault	1. Mechanical/weld/equipment failure
2. Hot tap	2. Operational	• System malfunction • Human error	2. Incorrect operation
3. Corrosion	3. Corrosion	• External • Internal • Stress cracking	3. Corrosion
4. Ground movement	4. Natural hazard	• Ground movement • Other	4. Natural force damage
5. External interference	5. Third party activity	• Accidental • Malicious • Incidental	5. Other outside force damage
			6. Excavation damage
6. Other/unknown			7. All other causes

TABLE 4.2

Parameters Considered for Gas Pipeline Failure – EGIG

SI. No.	Main Cause	Parameters
1	Third party damage	Diameter of pipeline, depth of cover, wall thickness
2	Corrosion	Year of construction, type of coating, wall thickness
3	Construction defect/material failure	Year of construction
4	Natural hazard	Diameter of pipeline
5	Others	Main causes
6	Hot tap error	Diameter of pipeline

4.2 Mitigation Measures

It is important to note that the reports do not specifically identify the type of mitigation measures employed in the pipelines from where the data originated. For example, EGIG report (European Gas Pipeline Incident Data Group, 2014) contains analysis of the parameters as shown in Table 4.2.

CONCAWE (Davis, Diaz, Gambardella, and Uhlig, 2013) for oil pipelines analyzes the sub-causes further as given in Table 4.3.

TABLE 4.3

CONCAWE Report: Analysis of Sub-Causes

Main Causes	Sub Causes
1. Mechanical failure	• Design [Incorrect design] • Construction [Faulty weld, Construction damage, incorrect installation • Materials fault [Incorrect material specification]
2. Operational	• System malfunction [Equipment, Instrumentation & control systems] • Human error [Incorrect operations, maintenance, procedures]
3. Corrosion	• External • Internal • Stress cracking
4. Natural hazard	• Ground movement [Landslide, subsidence, earthquake, flooding] • Other
5. Third party activity	• Accidental [Drilling/blasting, bulldozing, digging/trenching] • Malicious • Incidental

In order to develop Bayesian Network (BN), influencing factors are needed and therefore the nature of mitigation measures was taken from Muhlbauer (2004); Unnikrishnan, Shrihari, and Siddiqui (2015b); Pettitt, and Morgan (2009); Health and Safety Executive (2016); Tuft (2014) and from experts' opinion. Based on the above, a list of causes and sub-causes and mitigation measures adopted to counter the causes for pipeline failures were finalized. They are given in Table 4.4. Note that ME in the last column stands for Manual Entry.

Note 1: Mitigation measures are expressed as 'Not available or not done' in most cases to match with the syntax usage of NoisyOr Distribution in the BN. An illustration of how non-availability of the mitigation measures is combined for causation of the System malfunction (Under serial number 2 in Table 4.4) using NoisyOR distribution is given in Figure 4.2.

A similar methodology has been adopted for the causation of Human error and Failure due to corrosion. This is further discussed in Section 4.4.

It is also noted that there is lack of data on how much each of the mitigation measures have reduced pipeline failure rates.

4.3 BN for Loss of Containment from Pipeline

Based on the above causes, effects and mitigation measures and their inter-relationships, BN has been developed for loss of containment (LOC) from pipeline and is given in Figure 4.3. Each of the main causes have been

TABLE 4.4

Causes & Mitigation Measures for the Prevention of Pipeline Failures

Sl. No.	Main Cause (Node Name)	Sub-Cause (Node Name)	Mitigation Measure (Note 1)	States
1	Construction defect/mechanical failure (ConstDefectMatFailure)			Yes No (ME)
		Construction failure (ConstFailure)	Procedures and implementation	None, Average, Good (ME)
			Supervision	Adequate, Not adequate (ME)
		Defective design/materials fault (DefectivedesignOrMat)	Design factors	Yes, No (ME)
			Procedures & review not adequate	Yes, No (ME)
			Intelligent pigging not available	Yes, No (ME)
2	Operational failure	System malfunction (SystemMalfunction)		Yes, No (ME) Yes, No (Equation) NoisyOrDist (SystemMalfunction, 0.001, SCADANotAvailable, 0.05, OverPrProtectionNotAvail, 0.15, SafetySystemsHIPPSNotAvail, 0.10, HazardIdentificationNotDone, 0.10, RiskAssessmentNotDone, 0.10, CompositionMonitoringNotDone, 0.10, MOCProceduresNotAvail, 0.10 (See Section 5.4)

(Continued)

TABLE 4.4 (*Continued*)

Causes & Mitigation Measures for the Prevention of Pipeline Failures

Sl. No.	Main Cause (Node Name)	Sub-Cause (Node Name)	Mitigation Measure (Note 1)	States
			Supervisory Control and Data Acquisition (SCADA) not available	Yes, No (ME)
			Overpressure protection not available	Yes, No (ME)
			Safety systems (HIPPS) not available	Yes, No (ME)
			Hazard identification not done	Yes, No (ME)
			Risk assessments not done	Yes, No (ME)
			Composition monitoring not done	Yes, No (ME)
			Management Of Change (MOC) Procedure not available	Yes, No (ME)
		Human error (HumanError)		Yes, No (Equation) NoisyOrDist (HumanError, 0.005, TrainingNotAdequate, 0.20, OpAndMManualNotAvailNotR, 0.10, DrawingsNotUpToDate, 0.15, SafetyCultureNotPositive, 0.10)
			Training not adequate	Yes, No (ME)
			Operations & Maintenance manual not available or not reviewed	Yes, No (ME)
			Drawings not up-to-date	Yes, No (ME)
			Safety culture not positive	Yes, No (ME)

(*Continued*)

TABLE 4.4 (*Continued*)

Causes & Mitigation Measures for the Prevention of Pipeline Failures

Sl. No.	Main Cause (Node Name)	Sub-Cause (Node Name)	Mitigation Measure (Note 1)	States
3	Failure due to corrosion (FailureDueToCorrosion)			Yes, No (Equation) NoisyOrDist (FailureDueToCorrosion, 0.002, InternalCorrosion, 0.30, ExternalCorrosion, 0.25, DetectionOfSCC, 0.25, IntelligentPiggingNotAvail, 0.20)
			Intelligent pigging not available	Yes, No (ME)
		External corrosion (ExternalCorrosion)	Cathodic protection not available	Yes, No (ME)
			Pipeline coating not available	Yes, No (ME)
		Internal Corrosion (InternalCorrosion)	Internal lining not available	Yes, No (ME)
			Corrosion inhibitor Inj. not available	Yes, No (ME)
			Fluid corrosivity not considered	Yes, No (ME)
		Sulfide stress cracking (DetectionOfSCC)	Closed interval survey not done	Yes, No (ME)
4	Failure due to natural hazard (FailureDueToNatuaralHazards)	Ground movement/subsidence (Subsidence)		Yes, No (ME)
		Flooding (Flooding)		Yes, No (ME)

(Continued)

TABLE 4.4 (*Continued*)

Causes & Mitigation Measures for the Prevention of Pipeline Failures

Sl. No.	Main Cause (Node Name)	Sub-Cause (Node Name)	Mitigation Measure (Note 1)	States
5	Third party activity (FailureDueToThirdPartyactivity)	Other (Other)		Yes, No (ME)
		Accidental	Increase in wall thickness not adequate	Yes, No (ME) Equation. Please see section 5.4
		Malicious	Pipeline safety zones not identified	Yes, No (ME)
		Incidental	Depth of cover minimum 1M not provided	
			Warning marker posts not available	
			Plastic marker tapes not installed	
			Concrete slabbing not provided	
			Physical barriers not provided	
			Vibration detection not available	
			Right Of Way patrolling not done	
			Video cam monitoring not available	
			Site survey before construction not done	
6	Failure due to other causes (FailureDueToOtherCauses)			Yes, No (ME)

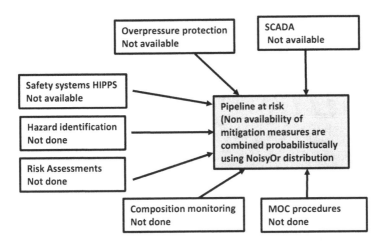

FIGURE 4.2
Illustration of NoisyOR distribution showing status of mitigation measures.

modeled with the sub-causes as parent nodes. The parents have binary states except node 'Procedures and implementation', which has three states 'None, Average, Good'. The parent nodes have been formulated as probability of not implementing a mitigation measure to match the syntax of the network. For example, for the factor, 'Failure due to third party activity', the mitigation measure of providing 1 m depth of cover is formulated as node 'DepthOfCover1MNotProvided' with binary states yes and no.

As can be seen from the BN in Figure 4.3, 50% probability has been assumed instead of a 100% negative state, for the parent nodes 'Supervision', 'Procedure And Reviews Not Adequate', 'Training Not Adequate', 'Cathodic Protection Not Available' and 'Plastic Marker Tape Not Installed' in line with general industry practice.

It is worthwhile to note that as the number of parent nodes increases, the number of entries in the Conditional Probability Table (CPT) goes up. For example, there are 11 parents for child node 'Failure due to third party activity'. Therefore with two states for each parent there will be $(2^{11}) \times 11$ entries in the CPT. In such a situation NoisyOr distribution is used to reflect the contribution of each parent.

4.4 NoisyOr Distribution

NoisyOr distribution can be used when there are several possible causes for an event, any of which can cause the event by itself, but only with a certain probability. Also, the event can occur spontaneously (without any of the

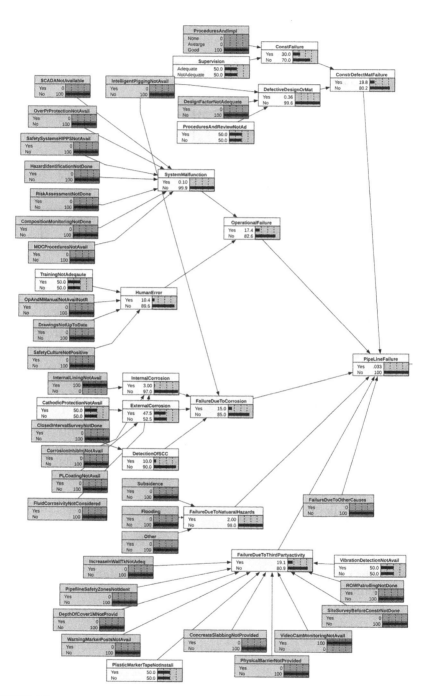

FIGURE 4.3
BN for gas pipeline LOC using NoisyOr.

known causes being true), which can be modeled with 'leak' probability. (This can be zero if it cannot occur spontaneously).

$$\text{NoisyOrDist}\left(e, \text{leak}, b1, p1,... bn, pn\right) \tag{4.1}$$

where e is the effect node, 'leak' is the leak factor which is the probability of the effect node even when all causes are zero, b1 is the node name for the cause and p1 is the probability of that cause impacting the effect node. The above can be written as Equation 4.2 for a better understanding.

$$\left(\text{Effect} \mid \text{Cause1, Cause2}\right) = \text{NoisyOrDist}\left(\text{Effect, leak, Cause1, p1, Cause2, p2}\right) \tag{4.2}$$

Application of the NoisyOr distribution is given below with regard to the node 'FailureDueToThirdPartyactivity'. Refer Figure 4.4.

For filling the probability values in the CPT for the node 'FailureDueToThirdPartyactivity', the NoisyOr distribution equation is written as in Equation 4.3. Further details are available in Netica (2017).

FailureDueToThirdPartyactivity | Increase In Wall Tk Not Adeq,

Pipelline Safety Zones Not Ident, Depth Of Cover1M Not Provid,

Warning Marker Posts Not Avail, Plastic Marker Tape Not Install,

Concreate Slabbing Not Provided, Physical Barrier Not Provided,

Vibration Detection Not Avail, ROW Patrolling Not Done, Video Cam

Monitoring Not Avail, Site Survey Before Constr Not Done) =

NoisyOrDist (FailureDueToThirdPartyactivity, 0.0035, Increase In Wall Tk

Not Adeq, 0.30, Pipelline Safety Zones Not Ident, 0.10, Depth Of Cover1M

Not Provid, 0.35, Warning Marker Posts Not Avail, 0.35, Plastic Marker

Tape Not Install, 0.20, Concreate Slabbing Not Provided, 0.20,

Physical Barrier Not Provided, 0.20, Vibration Detection Not Avail,

0.10, ROW Patrolling Not Done, 0.10, Video Cam Monitoring Not Avail,

0.05, Site Survey Before Constr Not Done, 0.20) (4.3)

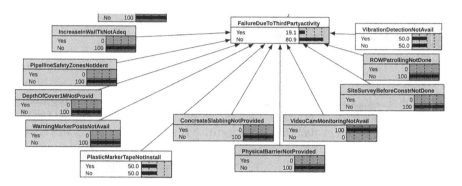

FIGURE 4.4
BN for sub-cause failure due to third party activity.

In the above equation, the first value on the RHS after the 'Failure Due To Third Party activity' is the leak factor that determines the probability of the event, even if all causes are not true. This has been set to 0.0035. The rest of the values after each cause (parent) node name represents the probability of that particular cause (parent) impacting the effect (child) node. Once the NoisyOr distribution is specified the CPT in table is automatically filled by the software with the probability values. A portion of the CPT for the node 'FailureDueToThirdPartyactivity' is shown in Table 4.5.

Parent nodes for all the six main causal factors have been formulated on the above basis. It is emphasized that the scale of probabilities for each of the main casual factors is independent of other causes. The resulting node Pipeline failure has been given the NosiyOr distribution to combine all the main causal factors to yield a generic failure rate matching with the current EGIG gas pipeline failure rate. This is given in Equation 4.4.

P (PipeLineFailure | FailureDueToCorrosion, FailureDueToNatuaralHazards,

FailureDueToThirdPartyactivity, OperationalFailure, ConstrDefectMatFailure,

FailureDueToOtherCauses)=

NoisyOrDist (PipeLineFailure, 0.00018, FailureDueToCorrosion, 0.000265,

FailureDueToNatuaralHazards, 0.00017, FailureDueToThirdPartyactivity,

0.00030, OperationalFailure, 0.00006, ConstrDefectMatFailure, 0.000175,

FailureDueToOtherCauses, 0.000080)   (4.4)

Once the BN has been set up it can be analyzed for impact of findings on any node/s on the pipeline failure and LOC scenario.

TABLE 4.5

Portion of CPT for the Node 'FailureDueToThirdPartyActivity'

FailureDueToThirdPartyactivity: R1-14-10-15

Yes	No	Increase InWallTk NotAdeq	Pipeline Safety Zones NotIdent	DepthOf Cover1M NotProvid	Warning Marker Posts NotAvail	Plastic Marker Tape NotInstall	Concrete Slabbing Not Provided	Physical Barrier Not Provided	Vibration Detection NotAvail	ROW Patrolling NotDone	VideoCam Monitoring NotAvail	SiteSurvey Before Constr NotDone
0.916399	0.083601	Yes	Yes	Yes	Yes	Yes	Yes	Yes	Yes	Yes	Yes	Yes
0.895498	0.104502	Yes	Yes	Yes	Yes	Yes	Yes	Yes	Yes	Yes	Yes	No
0.911999	0.088001	Yes	Yes	Yes	Yes	Yes	Yes	Yes	Yes	Yes	No	Yes
0.889998	0.110002	Yes	Yes	Yes	Yes	Yes	Yes	Yes	Yes	Yes	No	No
0.90711	0.09289	Yes	Yes	Yes	Yes	Yes	Yes	Yes	Yes	No	Yes	No
0.883887	0.116113	Yes	Yes	Yes	Yes	Yes	Yes	Yes	Yes	No	No	No
0.902221	0.097779	Yes	Yes	Yes	Yes	Yes	Yes	Yes	Yes	No	No	Yes
0.877776	0.122224	Yes	Yes	Yes	Yes	Yes	Yes	Yes	Yes	No	Yes	No
0.90711	0.09289	Yes	Yes	Yes	Yes	Yes	Yes	Yes	No	Yes	Yes	Yes
0.883887	0.116113	Yes	Yes	Yes	Yes	Yes	Yes	Yes	No	Yes	Yes	No
0.902221	0.097779	Yes	Yes	Yes	Yes	Yes	Yes	Yes	No	Yes	No	Yes
0.877776	0.122224	Yes	Yes	Yes	Yes	Yes	Yes	Yes	No	Yes	No	No
0.896788	0.103211	Yes	Yes	Yes	Yes	Yes	Yes	Yes	No	No	Yes	Yes
0.870986	0.129014	Yes	Yes	Yes	Yes	Yes	Yes	Yes	No	No	Yes	No
0.891356	0.108644	Yes	Yes	Yes	Yes	Yes	Yes	Yes	No	No	No	Yes
0.864195	0.135805	Yes	Yes	Yes	Yes	Yes	Yes	Yes	No	No	No	No
0.895498	0.104502	Yes	Yes	Yes	Yes	Yes	Yes	No	Yes	Yes	Yes	No
0.869373	0.130627	Yes	Yes	Yes	Yes	Yes	Yes	No	Yes	Yes	Yes	Yes
0.889998	0.110002	Yes	Yes	Yes	Yes	Yes	Yes	No	Yes	Yes	No	Yes
0.862498	0.137502	Yes	Yes	Yes	Yes	Yes	Yes	No	Yes	Yes	No	No

(Continued)

TABLE 4.5 (Continued)

Portion of CPT for the Node 'FailureDueToThirdPartyActivity'

FailureDueToThirdPartyactivity: R1-14-10-15

Yes	No	Increase InWallTk NotAdeq	Pipeline Safety Zones NotIdent	DepthOf CoverIM NotProvid	Warning Marker Posts NotAvail	Plastic Marker Tape NotInstall	Concrete Slabbing Not Provided	Physical Barrier Not Provided	Vibration Detection NotAvail	ROW Patrolling NotDone	VideoCam Monitoring NotAvail	SiteSurvey Before Constr NotDone
0.883887	0.116113	Yes	Yes	Yes	Yes	Yes	Yes	No	Yes	No	Yes	Yes
0.854859	0.145141	Yes	Yes	Yes	Yes	Yes	Yes	No	Yes	No	Yes	No
0.877776	0.122224	Yes	Yes	Yes	Yes	Yes	Yes	No	Yes	No	No	Yes
0.84722	0.15278	Yes	Yes	Yes	Yes	Yes	Yes	No	Yes	No	No	No
0.883887	0.116113	Yes	Yes	Yes	Yes	Yes	Yes	No	No	Yes	Yes	Yes
0.854859	0.145141	Yes	Yes	Yes	Yes	Yes	Yes	No	No	Yes	Yes	No
0.877776	0.122224	Yes	Yes	Yes	Yes	Yes	Yes	No	No	Yes	No	Yes
0.84722	0.15278	Yes	Yes	Yes	Yes	Yes	Yes	No	No	Yes	No	No
0.870986	0.129014	Yes	Yes	Yes	Yes	Yes	Yes	No	No	No	Yes	Yes
0.838732	0.161268	Yes	Yes	Yes	Yes	Yes	Yes	No	No	No	Yes	No
0.864195	0.135805	Yes	Yes	Yes	Yes	Yes	Yes	No	No	No	No	Yes
0.830244	0.169756	Yes	Yes	Yes	Yes	Yes	Yes	No	No	No	No	No
0.895498	0.104502	Yes	Yes	Yes	Yes	Yes	No	Yes	Yes	Yes	Yes	Yes
0.869373	0.130627	Yes	Yes	Yes	Yes	Yes	No	Yes	Yes	Yes	Yes	No
0.889998	0.110002	Yes	Yes	Yes	Yes	Yes	No	Yes	Yes	Yes	No	Yes
0.862498	0.137502	Yes	Yes	Yes	Yes	Yes	No	Yes	Yes	Yes	No	No
0.883887	0.116113	Yes	Yes	Yes	Yes	Yes	No	Yes	Yes	No	Yes	Yes
0.854859	0.145141	Yes	Yes	Yes	Yes	Yes	No	Yes	Yes	No	Yes	No
0.877776	0.122224	Yes	Yes	Yes	Yes	Yes	No	Yes	Yes	No	No	Yes
0.84722	0.15278	Yes	Yes	Yes	Yes	Yes	No	Yes	Yes	No	No	No

(Continued)

TABLE 4.5 (*Continued*)

Portion of CPT for the Node 'FailureDueToThirdPartyActivity'

FailureDueToThirdPartyactivity: R1-14-10-15

Yes	No	Increase InWallTk NotAdeq	Pipeline Safety Zones NotIdent	DepthOf Cover1M NotProvid	Warning Marker Posts NotAvail	Plastic Marker Tape NotInstall	Concrete Slabbing Not Provided	Physical Barrier Not Provided	Vibration Detection NotAvail	ROW Patrolling NotDone	VideoCam Monitoring NotAvail	SiteSurvey Before Constr NotDone
0.883887	0.116113	Yes	Yes	Yes	Yes	Yes	No	Yes	No	Yes	Yes	Yes
0.854859	0.145141	Yes	Yes	Yes	Yes	Yes	No	Yes	No	Yes	Yes	No
0.877776	0.122224	Yes	Yes	Yes	Yes	Yes	No	Yes	No	Yes	No	Yes
0.84722	0.15278	Yes	Yes	Yes	Yes	Yes	No	Yes	No	Yes	No	No
0.870986	0.129014	Yes	Yes	Yes	Yes	Yes	No	Yes	No	No	Yes	Yes
0.838732	0.161268	Yes	Yes	Yes	Yes	Yes	No	Yes	No	No	Yes	No
0.864195	0.135805	Yes	Yes	Yes	Yes	Yes	No	Yes	No	No	No	Yes
0.830244	0.169756	Yes	Yes	Yes	Yes	Yes	No	Yes	No	No	No	No
0.869373	0.130627	Yes	Yes	Yes	Yes	Yes	No	No	Yes	Yes	Yes	Yes
0.836716	0.163284	Yes	Yes	Yes	Yes	Yes	No	No	Yes	Yes	Yes	No
0.862498	0.137502	Yes	Yes	Yes	Yes	Yes	No	No	Yes	Yes	No	Yes
0.828122	0.171878	Yes	Yes	Yes	Yes	Yes	No	No	Yes	Yes	No	No
0.854859	0.145141	Yes	Yes	Yes	Yes	Yes	No	No	Yes	No	Yes	Yes

As an example, let us see what will be the major contributing factors when there is a confirmed pipeline failure (LOC). When the node state for Pipeline failure is set to 'Yes = 100%' for the given set of prior conditions, the BN will recalculate the node values. It is seen that the main contributions factors are 'Failure due to third party activity' followed by 'Construction defect/material failure' and 'Failure due to Corrosion'. This is shown in Figure 4.5.

4.5 Sensitivities

Apart from the above, sensitivity of other nodes to the query node also can be found out from the network. All BN software has the feature of analyzing the network for sensitivity of one node to another node (see Section 4.4). For Netica®, this involves selecting the target node and then the parent nodes for which the sensitivity has to be analyzed. For the given situation, the sensitivity analysis of other nodes to Pipeline failure node is given in Figure 4.6.

As can be seen, amongst the various causes, the highest influencing factor in pipeline LOC is third party interference.

4.6 Event Tree for Pipeline LOC

The eventual consequences of a natural gas pipeline LOC will be jet fire, Vapor Cloud Explosion (VCE) or flashfire back to source or no ignition (toxic gas dispersion) or a combination of the above. Main safety barriers preventing escalation of a LOC to the above scenarios are gas detection and Emergency Shutdown (ESD) actions. They are depicted in the BN in Figure 4.7. Since Netica®'s BN does not display more than four digit decimal, they have been separately generated and shown as text boxes in Figure 4.7. As can be seen from Figure 4.7, with the safety barriers in place and with current average value of pipeline failure (0.33 failures/10^3 km/year as per EGIG 2016 data [51]), the occurrence of hazardous consequences is quite low. The probability of failure is expressed as failures/1000 km/year.

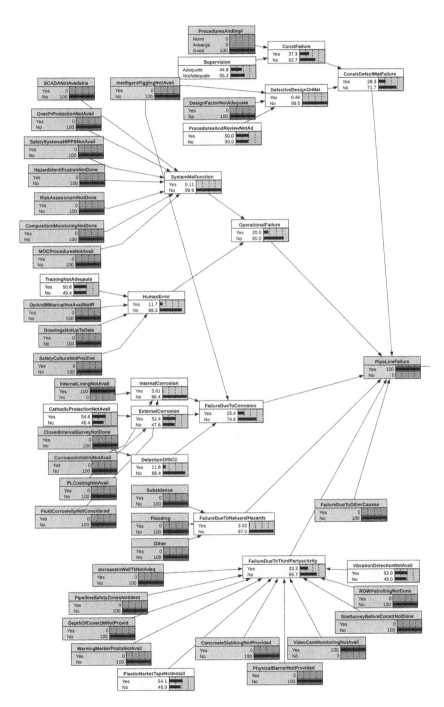

FIGURE 4.5
BN for confirmed LOC for pipeline.

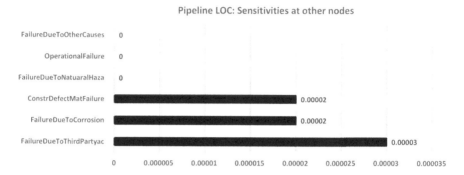

FIGURE 4.6
Sensitivity of LOC node to other nodes.

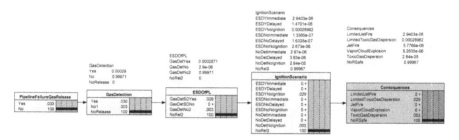

FIGURE 4.7
BN for consequences of pipeline LOC. Probability of failure is 0.033.

4.7 Case Study Using BN for Pipeline: Natural Gas Pipeline, Andhra Pradesh, India

4.7.1 Background

Gas Authority of India (GAIL) Tatipaka-Kondapalli 18 inch pipeline built in 2001 to carry natural gas from state-owned Oil and Natural Gas Commission (ONGC) wells to Lanco Power plant leaked sometime around 27 July 2014 near Nagaram village, East Godavari district in Andhra Pradesh, India leading to ignition and VCE. There were 22 fatalities, several injuries and considerable business loss. Government of India through ministry of oil constituted an internal enquiry. The inquiry committee's report is not available in the public domain; however only certain key findings of the same have been made known. The following is from the Press Trust of India in *The Economic Times* (Press Trust of India, 2014). According to press reports, the main causes of the accident have been noted as 'lack of systems approach'.

4.7.2 Key Findings

- The pipeline is supposed to carry dry natural gas. Yet there was no equipment (separators) to take out liquids in the line.
- The line had several leaks earlier which were controlled by temporary measure of clamps.
- The line was corroded and there were recommendations for injecting corrosion inhibitor, which was not done.
- People had reported smell of gas earlier several times, which went unheeded by the authorities.
- The report concluded 'inadequate systems approach' for the accident.

TABLE 4.6

Details of Nodes & States for Case Study: Pipeline Failure at AP, India

Main Causal Factor	Parent Nodes (Sub-Causes)	States	Probability %	Notes
Construction defect/mat. failure				All state values are given based on general industry practice.
Construction failure	Procedures and implementation	None	0	
		Average	0	
		Good	100	
	Supervision	Adequate	100	
		Not adequate	0	
Defective design or material	Design factor not adequate	Yes	0	
		No	100	
	Procedures and review not adequate	Yes	100	In view of what happened, this assumption is valid.
		No	0	
	Intelligent pigging not available	Yes	50	Assumed in the absence of information
		No	50	
Operational failure				All state values are given based on general industry practice.
System malfunction	SCADA not available	Yes	50	Assumed in the absence of information
		No	50	

(Continued)

TABLE 4.6 (*Continued*)

Details of Nodes & States for Case Study: Pipeline Failure at AP, India

Main Causal Factor	Parent Nodes (Sub-Causes)	States	Probability %	Notes
	Over Pr. protection not available	Yes	0	
		No	100	
	Safety system High Integrity Pressure Protection System (HIPPS) not available	Yes	0	
		No	100	
	Hazard identification not done	Yes	100	
		No	0	
	Risk assessment not done	Yes	100	
		No	0	
	Composition monitoring not done	Yes	100	
		No	0	
	MOC procedure not available	Yes	100	
		No	0	
Human error	Training not adequate	Yes	100	
		No	0	
	Op & Maint. manual not available or reviewed	Yes	100	
		No	0	
	Drawings not up to date	Yes	50	Assumed in the absence of information
		No	50	
	Safety culture not positive	Yes	100	
		No		
Failure due to corrosion				All state values are given based on general industry practice.

(*Continued*)

TABLE 4.6 (*Continued*)

Details of Nodes & States for Case Study: Pipeline Failure at AP, India

Main Causal Factor	Parent Nodes (Sub-Causes)	States	Probability %	Notes
Internal corrosion	Internal lining not available	Yes	100	
		No	0	
	Corrosion inhibitor Inj. not available	Yes	100	
		No	0	
	Fluid corrosivity not considered	Yes	100	
		No	0	
External corrosion	Cathodic protection not available	Yes	50	Assumed in the absence of information
		No	50	
	Pipeline coating not available	Yes	0	
		No	100	
Detection of Stress *corrosion* cracking (SCC)	Closed interval survey not done	Yes	50	Assumed in the absence of information
		No	50	
Failure due to third party activity				All state values are given based on general industry practice.
	Increase in wall thickness not adequate	Yes	0	
		No	100	
	Pipeline safety zones not adequate	Yes	100	
		No	0	
	Depth of cover 1 M not available	Yes	0	
		No	100	
	Warning marker posts not avail	Yes	50	
		No	50	
	Plastic marker tape not installed	Yes	0	

(*Continued*)

TABLE 4.6 (*Continued*)

Details of Nodes & States for Case Study: Pipeline Failure at AP, India

Main Causal Factor	Parent Nodes (Sub-Causes)	States	Probability %	Notes
		No	100	
	Concreate slabbing not provided	Yes	100	
		No		
	Physical barrier not provided	Yes	50	Assumed in the absence of information
		No	50	
	VideoCam monitoring not provided	Yes	100	
		No		
	Site survey before construction not done	Yes	50	Assumed in the absence of information
		No	50	
	Right of Way (ROW) patrolling not done	Yes	100	
		No	0	
	Vibration detection not available	Yes	100	
		No	0	
Failure due to natural hazards	Subsidence	Yes	0	All state values are given based on general industry practice.
		No	100	
	Flooding	Yes	0	
		No	100	
	Other	Yes	0	
		No	100	
Failure due to other causes		Yes	0	All state values are given based on general industry practice.
		No	100	

4.7.3 Application of the BN Model

The BN pipeline model can be tuned to see the situation by conducting a hindsight review of the impact of parent nodes on other parameters. The node states have been given the values that are thought to be the most likely situation before the LOC of the pipeline. They are given in Table 4.6 along with notes. All the values can be changed based on the information about actual situation when it is known.

Readers are encouraged to study the status of each of the parameters carefully to understand the implications of each of them in the risk profile of the pipeline.

4.7.4 BN for the Case Study

When the node state values in the BN are given the inputs as in Table 4.6 and simulated, it is seen that the percentage probability of pipeline failure has increased to 0.54/1000 km year, which is considerably higher than the current state in the industry. Please see Figure 4.8. A comparison of the values is given in Table 4.7 to illustrate this point.

With the above probability of LOC, the chance of a jet fire or VCE is still low (9.45×10^{-6} for jet fire and 8.61×10^{-6} VCE), provided the safety barrier, that is, gas detection is in operation. The ignition probabilities are based on International Association of Oil & Gas Producers data (IOGP, 2010). Please see Figure 4.9 for post-release LOC scenario with safety barriers in operation.

However, when LOC coincides with the failure of the key safety barrier for detection of the gas release, it can be seen that the probability of a fire is very high.

The probability goes up to 0.0175 for jet fire and 0.0159 for VCE from the average of 9.45×10^{-6} for jet fire and 8.61×10^{-6} VCE, respectively, when the detection barrier is not working (Figure 4.10). Thus, failure of gas detection in time played a major part in amplifying the incident to a major accident.

Clearly, the pipeline was operating in a high-risk situation. Had this been noticed in time, the accident could have been avoidable.

4.8 Chapter Summary

This chapter presents the BN for causes, mitigation measures employed to counter those causes, its interrelationships and post-event scenarios for LOC of hydrocarbon pipeline. Pipeline failure data are available in public domain and the same have been analyzed to find out the main causes and sub-causes of LOC. Further, the measures employed to mitigate these causes are investigated. It is noted that data regarding the same are mostly not available.

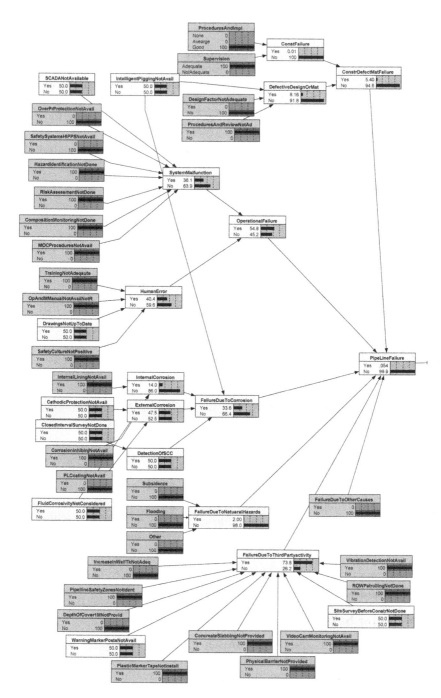

FIGURE 4.8
BN for GAIL pipeline with conditions before the accident.

TABLE 4.7

Comparison of GAIL Pipeline State with Industry Averages

Parameter	GAIL Pipeline	Industry Average (EGIG)
Failure frequency	$0.54/10^3$ km year	$0.33/10^3$ km year

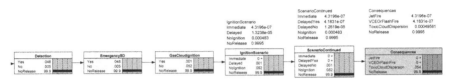

FIGURE 4.9
BN for GAIL pipeline consequences of LOC with safety barriers in operation.

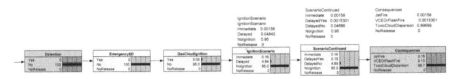

FIGURE 4.10
BN for Gail pipeline consequences of LOC-gas detection failure.

Industry practice and expert opinion have been incorporated for the above. The causes, sub-causes and mitigation measures have been converted to BN for LOC of the pipeline. The usage of NoisyOR distribution is described. Sensitivities and BN for the main causes are also given. A case study of a natural gas pipeline failure that happened in Andhra Pradesh, India has been included to illustrate the predictive nature of the BN.

Next chapter takes up BN for hazards of hydrocarbon storage tanks, namely Floating and Cone roof tanks.

5

Bayesian Network for Loss of Containment from Hydrocarbon Storage Tank

5.1 Storage Tank Basics

Large inventory of hydrocarbon liquids is usually stored in atmospheric storage tanks. In fact, a group of such tanks poses a high level of risk. Over the years, the design, operation and maintenance of atmospheric storage tanks have considerably improved. However, accidents like Buncefield (Buncefield Major Incident Investigation Board, 2007, 2008) and Jaipur Tank farm (Lal Committee, 2011) still happen. Figure 5.1 shows a typical hydrocarbon atmospheric storage tanks known as Floating roof tank.

In the Floating roof tank the roof deck floats on the liquid. Sealing between the tank wall and the deck is achieved by providing rim seal-primary seal (made of flexible polyurethane or similar material) and an additional secondary seal between the pontoon portion of the deck and tank wall.

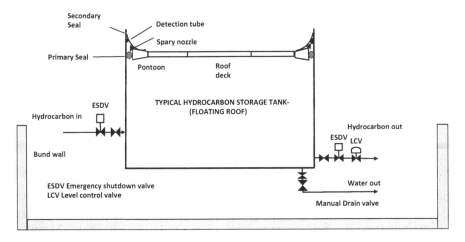

FIGURE 5.1
Schematic diagram of floating roof tank.

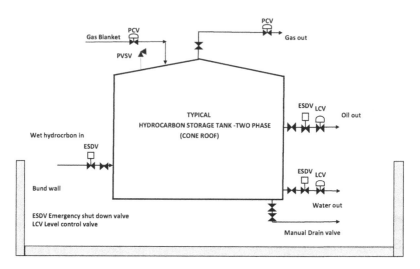

FIGURE 5.2
Schematic diagram of cone roof tank – two phase.

The protective safety barriers are the Emergency Shutdown Valve (ESDV) provided at the inlet and outlet of liquid lines, which are activated when a set of process conditions reaches predefined values. The predefined parameters include high-high and low-low levels inside the tank. Also, any fire starting on the top of the rim seal area is detected by a linear heat detector tube triggering an automatic foam outflow through the spray nozzles.

On the other hand, the Cone roof tank has a fixed conical roof and thus has a gas space above the liquid level. A gas blanket with control system is provided to ensure a positive pressure on the gas space. Figure 5.2 shows the schematic diagram of a Cone roof tank used to store two phases (oil & water) in oil Gathering Centers.

Protective devices include ESDVs at inlet and outlet lines that will act automatically to close on fulfilling a set of process conditions including high-high and low-low levels in the tank. The pressure vacuum safety valve (PVSV) provided on roof is the safety barrier to prevent over-pressurization as well as vacuum inside the tank.

5.2 Causal Factors for Loss of Containment

Since Bayesian network (BN) requires causal factors for loss of containment (LOC) from hydrocarbon storage tanks, the same was finalized on the basis of findings from the data given in LASTFIRE (2001); INERIS (2004); Flemish Government (2009); E & P Forum (1992); TNO-(VROM) (2005); Health and Safety Executive (2012); and International Association of Oil & Gas

TABLE 5.1

Key Causal Factors for Storage Tank Failures

Nodes for Key Causal Factors	No. of Parent Nodes
1. Quality of design	21
2. Quality of maintenance & inspection	8
3. Quality of construction	4
4. Quality of equipment selection	7
5. Quality of risk assessments	3
6. Quality of systems & procedures	12
7. Quality of human & organizational factors	7
8. Lightning strike	-
9. Catastrophic tank failure	3

Producers (2010). Accident investigation reports for Buncefield (Buncefield Major Incident Investigation Board, 2008) and Jaipur Tank farm accident (Lal Committee, 2011) also provided useful inputs. Key causal factors are grouped under nine headings and form the key nodes in the BN. They are from Atherton and Ash (2005); Chang and Lin (2006); Argyropoulos, Christolis, Nivolianitou and Markatos (2012); Huang and Mannan (2013); Necci, Argeni, Landucci, and Cozzani (2014); Ramnath (2013); Kang, Liang, Zhang, Lu, Liu and Yin (2014). The same is given in Table 5.1.

Each of the above causal nodes have been parameterized as Poor (0.3–0.5), Average (0.5–0.7) and Good (0.7–1).

Each of the key causal factors in Table 5.1 are child nodes of several parent nodes, which are the root causes influencing the above key causal nodes. Details of the parent nodes and its states are given in subsequent sections.

The relative probabilities of effects have been arrived at by combining the above key causes and intermediate causes suitably using NoisyOr distribution. The last effect node is arrived at by converting the relative probability to the generic average probability values given in the references given in Table 2.5 (Toegepast Natuurwetenschappelijk Onderzoek (Netherlands Organization for Applied Scientific Research (TNO (VROM), 2005).

5.3 Methodology for the Development of BN for LOC and Evaluation

The parent nodes of the child nodes in Table 5.1 have states with manual binary inputs effectively meaning 'Not fulfilling a requirement' 0% or 'Fulfilling a requirement' 100%. These are then combined using Normal

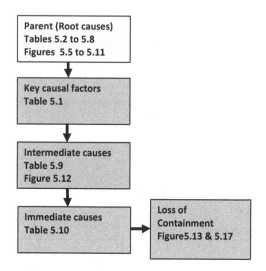

FIGURE 5.3
Overview of analysis of causal factors for LOC for storage tank.

distribution to bring the effects of the child nodes to probabilistic nature. The mean of the distribution is the average of all the parent nodes. Thus, if any of the parent node's state changes, it will be reflected in the state value of the child (effect) node.

An overview of the causal factors; one level preceding the key causal factors and three levels downstream of the key causal factors are shown in Figure 5.3 along with the tables and figures where data on the same are detailed.

BN has been developed separately for Floating roof and Cone roof tank, since there are certain differences in the design and operation of these tanks. Cone roof tanks as shown in Figure 5.2 do not have a rim seal. It has a gas space that is contained and pressurized with hydrocarbon gas to ensure positive pressure. PVSVs are fitted onto the roof as protection against overpressure as well as vacuum conditions inside the tank.

Event Trees and its equivalent BN have been developed separately for Floating and Cone roof tanks. An overview of the analysis is given in Figure 5.4.

The following gives a description of each of the key causal factors listed in Table 5.1, its parents (influencing factors) and the related BN.

5.3.1 Quality of Design

Table 5.2 describes this node, its parents (factors affecting this node) and its states along with the parameterization method. Readers are requested to study the above table in detail for a full appreciation of the causal factors.

The nodes in Table 5.2 and its cause and effect connectivities have been translated to a BN as shown in Figure 5.5.

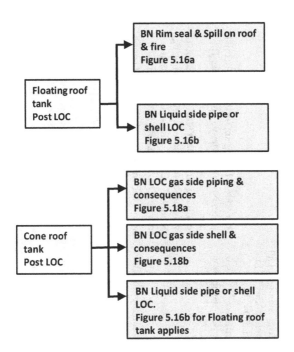

FIGURE 5.4
Overview of analysis of post LOC for floating and cone roof tanks.

As can be seen from Figure 5.5, all the 21 root causes have been linked to the key causal factor node 'Quality of design' (QOfD). Normal distribution at the node 'QOfD' (Quality of Design) combines the parent nodes as per the equation below:

P (QOfD | BDC, BIFR, CD, EStdsDesignChk, FireSVal, FGDS, CCTV,

FPSD, LPD, SD, SInS, MatSel, SelStTank, OFProtec, PipingFP,

TertiaryCont, FaultTD, ESDSIF, StatEl, TempMon, QOfSysProcScore,

ElectHazAreaClass) = NormalDist (QOfD, avg (BDC, BIFR, CD,

EStdsDesignChk, CCTV, FireSVal, FGDS, FPSD, LPD, SD, SInS,

MatSel, SelStTank, OFProtec, PipingFP, TertiaryCont, FaultTD, ESDSIF,

StatEl, TempMon, QOfSysProcScore, ElectHazAreaClass), 0.1)

(5.1)

Of the above, all sub-causes have binary states except 'EStdsDesignChk', which is defined by an equation using NoisyOr distribution given in Equation 5.2

TABLE 5.2

Details of the Parent Nodes of QOfD

Sl. No.	Main Node and Its Parent Nodes	Node Full Form	States	Parameterization Method	Description
1	**QOfD**	**Quality of design**	Poor Average Good	Calculated. Normal distribution. (Mean: average of all parent nodes 1.1 to 1.21. SD = 0.1)	This node is having six further parents with states No/Yes – Manual input node. They are:
1.1	EStdsDesignChk	Adherence to engineering standards and regulations & design checks	Part Full	Calculated NoisyOr distribution	Following all relevant Stds, Drains double valving, automatic tank level monitoring, Remote isolation valves, Anti-rotation device, Double seal for Floating roof.
1.2	OFProtection	Overflow protection	Adequate Not adequate	Manual. Input node	Automatic valve provided to cut off supply in case of overfill
1.3	SelStTank	Selection of storage tank type	Incorrect Correct	Manual. Input node	Selection of the type of storage, i.e. fixed or floating, is based on the Flash point of the liquid.
1.4	SInS	Site Inspection and study	Not adequate Adequate	Manual. Input node	Topographical and other relevant information about the site is critical.
1.5	SD	Safe distances	Not adequate Adequate	Manual. Input node	Best practices allow certain minimum distances between tanks to be fixed. However generally this is confirmed by the risk assessment.
1.6	BDC	Bunds/dikes capacities	Not adequate Adequate	Manual. Input node	Dike capacity is generally 1.25 times the tank capacity. But when there are more than one tank within a dike, the situation has to be analyzed on a case-to-case basis.

(Continued)

TABLE 5.2 (*Continued*)

Details of the Parent Nodes of QOfD

Sl. No.	Main Node and Its Parent Nodes	Node Full Form	States	Parameterization Method	Description
1.7	BFR	Bund fire resistance	Poor Good	Manual. Input node	Integrity of bund/dike wall which is the secondary containment is important and provision with valve for proper bund draining
1.8	FireSVal	Fire safe valves	Not provided Provided	Manual. Input node	Valves within bund wall shall be fire safe.
1.9	CD	Capacity definition of tank	Not clear Clear	Manual. Input node	There must be clarity in capacity definition of tanks capacity mentioned in the design documents. Usually it is the working capacity between the low level to high level control band.
1.10	LPD	Lightning protection design	Not adequate Adequate	Manual. Input node	
1.11	FPSD	Fire protection and fire protection system design	Not adequate Adequate	Manual. Input node	Including cooling water & foam systems for Floating (with dams) and Cone Roof tanks
1.12	PipingFP	Piping fire proofing	Not provided Provided	Manual. Input node	All piping within Bund wall shall be adequately fireproofed.
1.13	FGDS	Fire and gas detection system	Not provided Provided	Manual. Input node	Hydrocarbon detection.
1.14	CCTV	Closed circuit TV	Not provided Provided	Manual. Input node	Linear heat detection for rim seals.

(Continued)

TABLE 5.2 (*Continued*)
Details of the Parent Nodes of QOfD

Sl. No.	Main Node and Its Parent Nodes	Node Full Form	States	Parameterization Method	Description
1.15	StatEl	Static electricity prevention	Not considered / Considered	Manual. Input node	Velocity of all fluid movement into the tank, out of the tank and within the tank must be analyzed. Accumulation of static change must be prevented either by design or procedures, proper bonding & earthing.
1.16	MatSel	Material selection	Not adequate / Adequate	Manual. Input node	Correct metallurgy of the tank and components are essential. Usually internal fiberglass lining is also provided.
1.17	TempMon	Temperature monitoring	Not provided / Provided	Manual. Input node	
1.18	ESDSIF	Emergency Shut down valve as per safety instrumented function requirements	Not provided / Provided	Manual. Input node	ESDV valve shall be provided at the inlet and outlet of tanks and its Safety Integrity Level (SIL) level as per ISA 61511 must be analyzed by Layers Of Protection Analysis (LOPA).
1.19	ElectHazAreaClass	Electrical area classification	Not adequate / Adequate	Manual. Input node	Accurate electrical hazardous area classification is required.
1.20	TertiaryContainment	Tertiary containment	Not considered / Considered	Manual. Input node	Tertiary containment must be considered. In both Buncefield and Jaipur the secondary containment failed.
1.21	FaultTD	Fault tolerant design	Not considered / Considered	Manual. Input node	Automation of operations must be a prime objective with adequate backup and plan for human intervention. In Buncefield the automatic shut off valve did not work and in Jaipur the LOC that became uncontrollable during the transfer operations was the triggering point of the accident

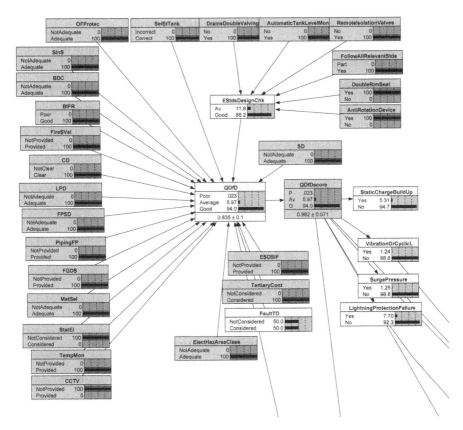

FIGURE 5.5
BN for the node quality of design.

P (EStdsDesignChk | DrainsDoubleValving, AutomaticTankLevelMon,

RemoteIsolationValves, DoubleSealForFR, AntiRotationDesignFR,

FollowAllRelevantStds) =

NoisyOrDist (EStdsDesignChk, 0.00001, DrainsDoubleValving, 0.20,

AutomaticTankLevelMon, 0.20, RemoteIsolationValves, 0.20,

DoubleSealForFR, 0.20, AntiRotationDesignFR,

0.20, FollowAllRelevantStds, 0.70)   (5.2)

Given the condition of the states for each of the parent nodes, the probability values for QOfD is calculated in percentages as Poor = 0.023, Average = 6.00, Good = 94.0 (Poor, Average and Good categorization is on the basis of the value of Normal distribution between 0.3–0.5, 0.5–0.7 and 0.7–1, respectively).

5.3.2 Quality of Maintenance and Inspection

Table 5.3 shows the parent nodes, its description and parameterization methods for the main causal factor node Quality of Maintenance and inspection. The corresponding BN in Figure 5.6.

Figure 5.6 shows the direct connection of the eight influencing factors (parent nodes) on the node 'Quality of Maintenance and Inspection' (QOfMaintInsp).

Equation for node 'Quality of Maintenance & inspection' is given below:

P (QOfMaintInsp | RI, PRFTest, PreVent, ProEq, ExPrEq, HotWr,

Training, WComChk, QOfSysProcScore) =

NormalDist (QOfMaintInsp, avg (RI, PRFTest, PreVent, ProEq,   (5.3)

ExPrEq, HotWr, Training, WComChk, QOfSysProcScore), 0.1)

TABLE 5.3

Details of Parent Nodes for Quality of Maintenance & Inspection

Sl. No.	Node Name	Node Full Form	States	Parameterization Method	Description
1	RI	Routine inspection	Yes No	Manual	Planned routine inspection
2	PRFTest	Proof testing	Yes No	Manual	Regular testing of components & systems including overfill protection system
3	PreVent	Preventive maintenance	Yes No	Manual	Scheduled preventive maintenance
4	ProEq	Protective equipment	Yes No	Manual	Usage of protective equipment during maintenance
5	ExPrEq	Explosion proof equipment	Yes No	Manual	Usage of explosion-proof equipment
6	HotWr	Hot work permit	Yes No	Manual	Established hot work permit system in tank area in accordance with relevant standards.
7	Training	Training	Yes No	Manual	Providing training for staff
8	WComChk	Work completion check	Yes No	Manual	A verification system for checking completion of work as per established procedures

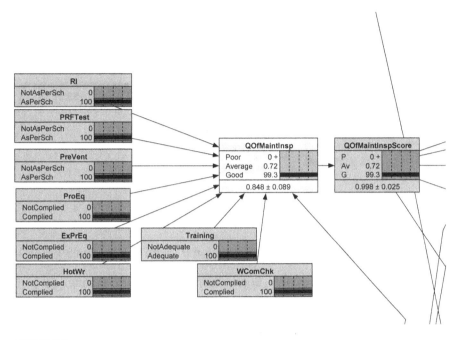

FIGURE 5.6
BN for the node quality of maintenance & inspection.

5.3.3 Quality of Construction

Table 5.4 gives a description for the parents influencing the main causal factor Quality of construction followed by BN in Figure 5.7.

TABLE 5.4

Details of Parent Nodes of Construction

Sl. No.	Node Name	Node Full Form	States	Parameterization Method	Description
1	ContrQuality	Contractor quality	Poor Average Good	Normal distribution	Quality of the contractor for executing the work
2	Supervision	Supervision	Not adequate Adequate	Manual	Adequacy of construction supervision
3	ConstCertPr	Construction certification procedures	Not implemented Implemented	Manual	Procedures for certifying completion of construction
4	TestingReq	Testing requirements	Not complete Complete	Manual	Completion of testing requirements

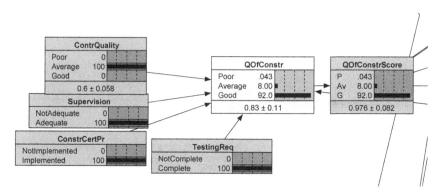

FIGURE 5.7
BN for the node quality of construction.

Equation for node 'Quality of construction is given in Equation 5.4

P (QOfConstr | ContrQuality, Supervision, ConstrCertPr, TestingReq,

QOfSysProc) =

NormalDist (QOfConstr, avg (ContrQuality, Supervision, ConstrCertPr,

TestingReq, QOfSysProc), 0.1)

$$(5.4)$$

5.3.4 Quality of Equipment Selection

Table 5.5 indicates the parent nodes for Quality of Equipment selection. Figure 5.8 shows the corresponding BN.

TABLE 5.5

Details of Parent Nodes of Quality of Equipment Selection

Sl. No.	Node Name	Node Full Form	States	Parameterization Method	Description
1	LiProtS	Lightning protection system	Not industry practice Industry practice	Manual	Lightning protection system selection should be as per industry practice
2	ESDVSys	Emergency Shut Down valve & system	Proven Not Proven	Manual	ESDV design alone is not enough; the system selected must be proven and fit for the purpose.

(Continued)

TABLE 5.5 (*Continued*)

Details of Parent Nodes of Quality of Equipment Selection

Sl. No.	Node Name	Node Full Form	States	Parameterization Method	Description
3	IsoVType	Isolation valve type	Not operator friendly Operator friendly	Manual	The isolation valve has a critical function. Its failure probability must be evaluated with respect to operator actions.
4	ROV	Remote operated valve	Not fire safe Fire safe	Manual	ROV provided must be fire safe as per relevant standards.
5	ComplianceWith ElecHaz	Compliance with electrical hazardous area classification	Not complied Complied	Manual	All equipment shall comply with appropriate electrical hazardous area classification
6	TherExForLiquid	Thermal expansion for liquid	Not provided Provided	Manual	If there is possibility of locked up liquid in above ground piping, then thermal expansion and suitable protection must be considered.
7	QfSysProc	Quality of Systems and procedures	Poor, Average, Good	Input from another calculated node	–

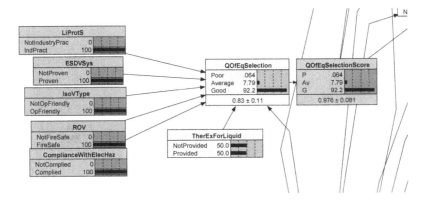

FIGURE 5.8
BN for the node quality of equipment selection.

Note: Figure 5.8 does not show the node 'QfSysProc' (Quality of Systems and Procedures) but shows only the connection from that node.

Equation for node 'Quality of 'Equipment selection' is given below:

P (QOfEqSelection | LiProtS, ESDVSys, IsoVType, ROV, TherExForLiquid,

QOfSysProcScore, ComplianceWithElecHaz) =

NormalDist (QOfEqSelection, avg (LiProtS, ESDVSys, IsoVType, ROV,

TherExForLiquid, QOfSysProcScore, ComplianceWithElecHaz), 0.1)

$$(5.5)$$

5.3.5 Quality of Risk Assessments

Table 5.6 gives the parent nodes and description for the main causal factor Quality of Risk assessments followed by BN in Figure 5.9.

TABLE 5.6

Details of Parent Nodes of Quality of Risk Assessments

Sl. No.	Node Name	Node Full Form	States	Parameterization Method	Description
1	HAZID	Hazard identification	Poor quality Good quality	Manual	Quality of hazard identification study
2	HAZOP	Hazard & operability studies	Poor quality Good quality	Manual	Quality of hazard & operability study
3	RA	Risk assessments	Poor quality Good quality	Manual	Quality of risk assessment studies
4	QOfSysProc	Quality of Systems and procedures	Poor, Average, Good	Input from another node	There must be evaluation of quality of risk assessments

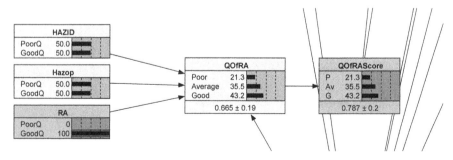

FIGURE 5.9
BN for the node for quality of risk assessments.

Node 'QOfRA' contains the equation 5.6

P (QOfRA | HAZID, RA, Hazop, QOfSysProcScore) =

NormalDist QOfRA, (0.24* HAZID + 0.24* RA + 0.24 * Hazop + 0.28   (5.6)

*QOfSysProcScore), 0.1)

Here the mean of the Normal distribution is the weighted average of the sub-causes (parents).

5.3.6 Quality of Systems and Procedures

Table 5.7 gives the parent nodes and description for the main causal factor Quality of construction followed by BN in Figure 5.10.
   Equation for the node 'QOSysProc' is given below:

P (QOfSysProc | EngControlsRev, SOPReview, EmergencyResponsPr,

EmergencyDrills, OpStaffTraining, AdherenceToMOC, AuditForPSM,

ProtocolForCRManning, DDUpdateSystem, PeriodicUpdateRA,

IsolationProcedures, DrainingProcedures, QOfHOF) =

NormalDist (QOfSysProc, avg (EngControlsRev, SOPReview,

EmergencyResponsPr, EmergencyDrills, OpStaffTraining, AdherenceToMOC,

AuditForPSM, ProtocolForCRManning, DDUpdateSystem, PeriodicUpdateRA,

IsolationProcedures, DrainingProcedures, QOfHOF), 0.1)

$$(5.7)$$

5.3.7 Quality of Human and Organizational Factors

Table 5.8 and Figure 5.11 give details of parent nodes for Quality Human and Organizational factors and the corresponding BN, respectively.
   Equation 5.8 provides the relationship between the parent and child nodes as Normal distribution with mean as the average of all parent nodes with a Standard deviation of 0.1.

P (QOfHOF | VisibilityPSM, UnderstandingPSM, KwPotRisks,

QualifiedPer, RRDefPSM, SelForSC, SafetyCulture) =

NormalDist (QOfHOF, avg (VisibilityPSM, UnderstandingPSM,

KwPotRisks, QualifiedPer, RRDefPSM, SelForSC, SafetyCulture), 0.1)

$$(5.8)$$

TABLE 5.7

Details of Parent Nodes of Quality of Systems and Procedures

Sl. No.	Node Name	Node Full Form	States	Parameterization Method	Description
1	EnggControls Rev	Engineering controls review	Not implemented Implemented	Manual	Review procedures for quality control of engineering work
2	SOPReview	Standard operating procedures review	Not available Available	Manual	Regular review of operating procedures
3	Emergency ResposneDrills	Emergency response drills	Not available Available	Manual	Periodic emergency response drills and review
4	OpStaffTraining	Operating staff training	Not conducted Conducted	Manual	Training program for operating staff
5	Adherence ToMOC	Adherence to management of change	Not complied Complied	Manual	Procedures for management of change and adherence to the same
6	AuditForPSM	Audit for process safety management	Not conducted Conducted.	Manual	Audit procedures and audit exercises on regular basis
7	ProtocolFor CRManning	Protocol for control room manning	Not available Available	Manual	Protocol for manning the control room
8	DDUUpdate System	Drawings and documentation update system	Not available Available	Manual	Ongoing system for updating drawings and documentation
9	Isolation Procedures	Isolation procedures	Not available Available	Manual	Established procedures for isolation and adherence to the same
10	Draining Procedures	Draining procedures	Not available Available	Manual	Established procedures for isolation and adherence to the same
11	Periodic UpdateRA	Periodic update of Risk assessments	Not complied Complied	Manual	Scheduled update of risk assessments
12	QOfHOF	Quality of human and organizational factors	Poor Average Good	From another node	

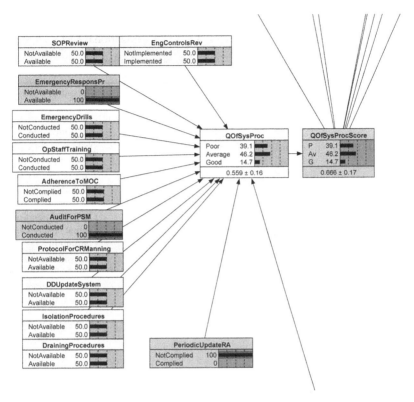

FIGURE 5.10
BN for the node quality of systems and procedures.

TABLE 5.8

Details of Parent Nodes of Quality of Human and Organizational Factors

Sl. No.	Node Name	Node Full Form	States	Parameterization Method	Description
1	VisibilityPSM	Visibility of Process Safety Management (PSM)	Not visible Visible	Manual	Process safety management must be visible to the staff for its effectiveness
2	Understanding PSM	Understanding PSM	Poor Good	Manual	PSM is often confused with personal and construction safety even at senior management levels.
3	KwPotRisks	Knowledge of potential risks	Poor Good	Manual	Risk awareness of personnel is important.

(Continued)

TABLE 5.8 (*Continued*)

Details of Parent Nodes of Quality of Human and Organizational Factors

Sl. No.	Node Name	Node Full Form	States	Parameterization Method	Description
4	QualifiedPer	Qualified personnel	Not available Available	Manual	Personnel must have the required technical qualifications
5	RRDefPSM	Roles and responsibility definitions for PSM	Not assigned Assigned	Manual	PSM hierarchy must have their roles and responsibility clearly defined
6	SelForSC	Selection of personnel for safety critical operations	Wrong choice Correct Choice	Manual	Personnel for safety critical operations must be assessed for suitability
7	Safety Culture	Safety culture	Pool Positive	Manual	A positive, risk aware safety culture is a basic requirement for promoting safe practices.

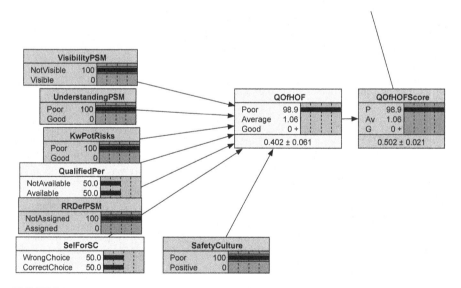

FIGURE 5.11

BN for the node quality of human and organizational factors.

5.3.8 Intermediate Causes

Each of the key causal factors further influence downstream intermediate causes which form the immediate causes for a failure. Certain key causal factors impact more than one intermediate factor. They are listed in Table 5.9.

TABLE 5.9

Intermediate Factors

Sl. No.	Key Causal Factor	Impacted Intermediate Downstream Factors
1	Quality of design	Static charge build up
		Vibration or cyclic loading
		Surge pressure
		Lighting protection failure
2	Quality of construction	Weld failure
		Tank internal lining failure
		Liquid side piping failure
3	Quality of maintenance and inspection	Tank outlet valve leak
	Quality of equipment selection	
		Liquid side piping failure
		Corrosion
		Liquid control system failure
		ESDV failure
4	Quality of Quantitative Risk Assessment (QRA)	Liquid side piping failure
		Excessive liquid transfer
5	Quality of systems and procedures	Incorrect valve operation
		Improper work on tank
		Tank outlet valve leak

5.3.9 Other Root Causes

They include lightning strike, catastrophic tank failure due to natural causes, foundation subsidence and external impact.

The intermediate and immediate causes and failure of components are given in Figure 5.12, which shows the relevant portion of the BN.

5.3.10 BN for LOC Scenarios from Floating Roof Tank

The nine key causal factors in Table 5.1, 63 root causes for the same, intermediate downstream causes plus the failure nodes given in Table 5.9 and their interrelationships are combined in a BN for different LOC scenarios for Floating roof tank. NoisyOr distribution or weighted average equations have been used for the effect (child) nodes to define the conditional probability tables (CPTs).

Full BN for LOC of Floating roof tank is given in Figure 5.13.

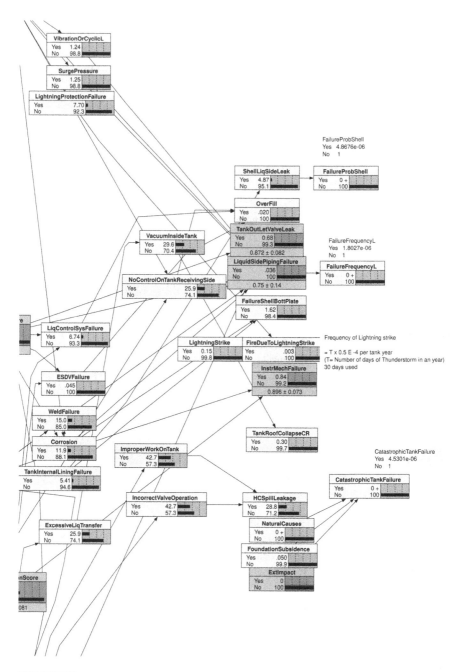

FIGURE 5.12
BN for the nodes' intermediate factors and immediate causes.

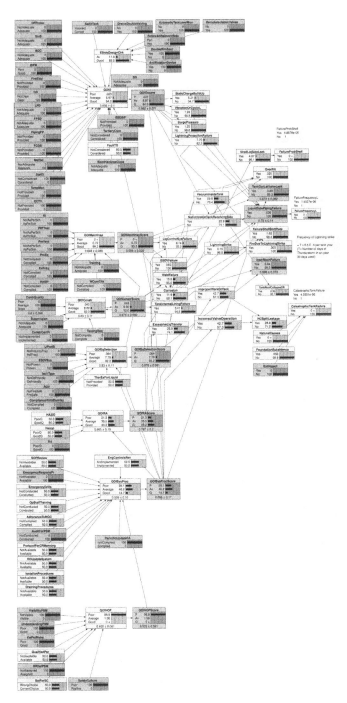

FIGURE 5.13
BN for LOC of floating roof tank.

The cause and effect relationships of the final, its predecessor (as applicable) and its immediate parent nodes are shown in BN in Figure 5.14. The CPT and notes of the same are given in Table 5.10.

These relative probabilities have been transformed to an average failure frequency (probabilities) based on the values indicated in the published literature (TNO (VROM), 2005) for LOC. Please see node FailureFrequencyL. The generic frequencies for LOC for atmospheric storage tank and for pipes are of the order of 5×10^{-6} and 2×10^{-6} per meter per year, respectively. Another source of data is International Association of Oil & Gas Producers Storage incident frequencies Report No. 434-03 (International Association of Oil & Gas Producers, 2010). This is done to provide a certain measure of judgment about the variation from current average when causal factors are changing.

5.3.11 Sensitivities

Sensitivity of other nodes to a target node can be analyzed in a BN. Figure 5.14 indicates sensitivity of other nodes to shell side leak, and Figure 5.15 shows sensitivity of other nodes to failure frequency (probability) of liquid side piping LOC.

Nodes with zero sensitivity have been omitted in both the figures to fit the bars to the scale.

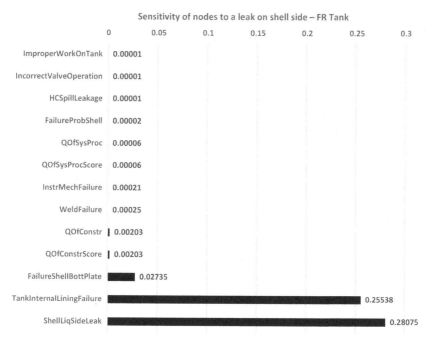

FIGURE 5.14
Sensitivity of nodes to a leak on shell side FR tank.

TABLE 5.10

CPT for Final Effect Nodes

Final Effect Node Name and its Parent Nodes in ()	CPT	Notes
FailureProbShell (ShellLiqSideLeak)	FailureProbShell: ShellLiqSideLeak Yes No Yes 1e-4 0.9999 No 0 1	The CPT converts the relative probability to average generic probability.
ShellLiqSideLeak (TankInternalLiningFailure)	ShellLiqSideLeak: TankInternalLiningFailure Yes No Yes 0.9 0.1 No 0 1	Probability of failure of tank internal lining directly affects the probability of shell leak.
OverFill (NoControlOnTankReceivingSide, LiqControlSysFailure, ESDVFailure)	p (OverFill | NoControlOnTankReceivingSide, LiqControlSysFailure, ESDVFailure) = NoisyOrDist (OverFill, 0.0000001, NoControlOnTankReceivingSide, 0.0005, LiqControlSysFailure, 0.001, ESDVFailure, 0.0020)	The relationship is defined by NoisyOr distribution.
TankOutLetValveLeak (QOfMaintInspScore, QOfSysProcScore)	TankOutLetValveLeak (QOfMaintInspScore, QOfSysProcScore) = 0.50* QOfMaintInspScore+0.50*QOfSysProcScore	The weighted equation defines the relationship of parents (causes) with effect.
FailureFrequencyL: (LiquidSidePipingFailure)	FailureFrequencyL: LiquidSidePiping Failure Yes No Yes 0.005 0.995 No 0 1	The CPT converts the relative probability to average generic probability.

(*Continued*)

TABLE 5.10 (Continued)

CPT for Final Effect Nodes

Final Effect Node Name and its Parent Nodes in ()	CPT	Notes
LiquidSidePipingFailure (Corrosion, SurgePressure, VibrationOrCyclicL, QOfMaintInspScore, QOfDscore, QOfConstrScore)	LiquidSidePipingFailure (Corrosion, SurgePressure, VibrationOrCyclicL, QOfMaintInspScore, QOfDscore, QOfConstrScore) = (0.35*Corrosion+0.05*SurgePressure+0.05*VibrationOrCyclicL+0.25*QOfDscore+0.20*QOfMaintInspScore+0.10*QOfConstrScore)	Liquid side piping failure is defined by NoisyOr distribution of its causes.

FailureShellBottPlate (TankInternalLiningFailure)

FailureShellBottPlate:

Yes	No	TankInternalLiningFailure	Corrosion
0.9	0.1	Yes	Yes
0.1	0.9	Yes	No
0.05	0.95	No	Yes
0	1	No	No

Notes: Failure of tank bottom plate is defined by the CPT of its parent nodes.

FireDueToLightningStrike (LightningProtectionFailure)

FireDueToLightningStrike:

Yes	No	LightningStrike	LightningProtectionFailure
0.1	0.9	Yes	Yes
0.01	0.99	Yes	No
0	1	No	Yes
0	1	No	No

Notes: CPT defines the relationship

(Continued)

TABLE 5.10 (*Continued*)

CPT for Final Effect Nodes

Final Effect Node Name and its Parent Nodes in ()	CPT	Notes

TankRoofCollapse (VacuumInsideTank)	TankRoofCollapse:
	
	               VacuumInsideTank
	
	              Yes
	              No
	
	    Yes No
	    0.01 0.99
	    0 1
		Tank roof collapse is directly dependent on the vacuum condition inside the tank

CatastrophicTankFailure (NaturalCauses, FoundationSubsidence, ExtImpact)	P (CatastrophicTankFailure | NaturalCauses, FoundationSubsidence, ExtImpact) = NoisyOrDist (CatastrophicTankFailure, 0.0000045, NaturalCauses, 0.000008, FoundationSubsidence, 0.00006, ExtImpact, 0.00001)	NoisyOr distribution is used to define the relationship between the parent and causal nodes

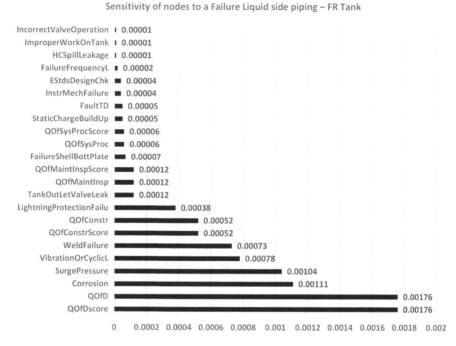

FIGURE 5.15
Sensitivity of nodes to a failure liquid side piping FR tank.

5.4 Event Tree for the Post LOC Scenario in Floating Roof (FR) Tank

BN equivalent of Event trees for the post LOC scenario for Floating roof tank are given below based on (LASTFIRE, 2001). The scenarios of LOC are

   i. Rim seal fire
  ii. Spill on roof
 iii. Shell or liquid side piping – short duration (small bund fire) or long duration (large bund fire)

The scenarios are shown in Figures 5.16a–c below.

Ignition probabilities are based on International Association of Oil & Gas Producers data (International Association of Oil & Gas Producers, 2010). The node names are self-explanatory. 'Failure of FFSystem' stands for failure of fire fighting system.

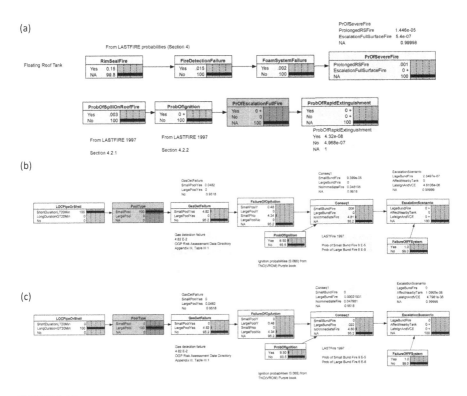

FIGURE 5.16
(a) BN for post LOC-rim seal leak or spill FR tank, (b) BN for post LOC-liquid side pipe or shell FR short duration, (c) BN for post LOC-liquid side pipe or shell FR long duration.

5.5 BN for LOC in Cone Roof (CR) Tank

As can be seen from Figure 5.2, Cone roof tank has a gas space, and blanketing gas keeps the space under positive pressure. Therefore, a node has been added to take care of this aspect. All other nodes, CPT and equations remain the same as in Floating roof tank. The BN for Cone roof tank is given in Figure 5.17.

The equation for the added node 'GasSidePipingFailure' is given in Equation 5.9

$$\begin{aligned}
&\text{GasSidePipingFailure (Corrosion, VibrationOrCyclicL, QOfDscore,} \\
&\text{QOfMaintInspScore, QOfConstrScore)} = (0.35* \text{ Corrosion} + 0.10* \\
&\text{VibrationOrCyclicL} + 0.25*\text{QOfDscore} + 0.10* \text{ QOfMaintInspScore} \\
&+0.10*\text{QOfConstrScore})
\end{aligned} \tag{5.9}$$

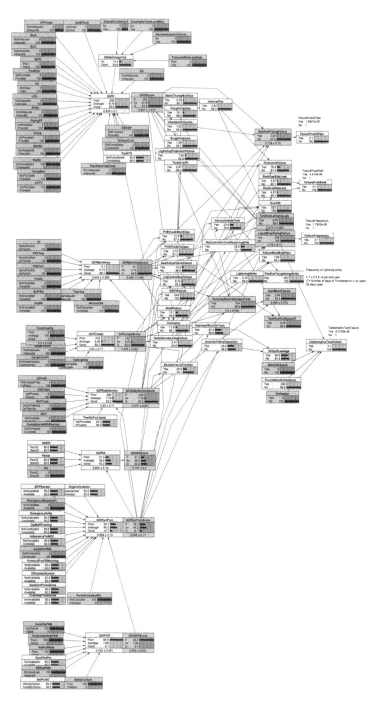

FIGURE 5.17
BN for cone roof tank generic.

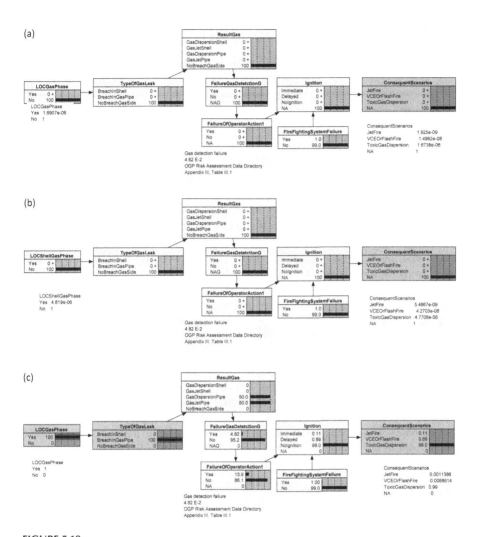

FIGURE 5.18

(a) BN for post LOC cone roof tank gas side piping leak, (b) BN for post LOC cone roof tank shell side gas leak, (c) BN for confirmed LOC from cone roof tank gas piping.

The next difference is in the LOC scenario. In Cone roof tank there is a probability of LOC for gas side either on the gas piping side or the shell side (gas). Please see top right-hand nodes in Figure 5.17. The BN equivalent of Event trees for LOC on gas side piping and shell side (gas) are given in Figure 5.18a and b. The probability of LOC is taken from Figure 5.17 (1.6907 e-06 for gas piping and 4.819 e-06 for shell side). As can be seen, with generic probabilities for LOC, the consequences are very low. However, when there is a confirmed LOC on the gas side piping (yes = 100%), the probability of a consequences, namely jet fire, Vapor Cloud Explosion (VCE) and toxic gas dispersion increases. This is shown in BN given in Figure 5.18c, with LOC being 100% on gas side piping.

5.6 Chapter Summary

The probability of LOC in the shell side (liquid) for Floating roof Tank and that on the Roof and Shell side for gas and liquid for Cone Roof tank have been described in detail in the preceding sections. The causal factors have been analyzed in four levels. Key causal factors and its root (parent) causes, followed by downstream effects of the key causal factors, namely intermediate, immediate causes and failure of components. The BN showing the inter-relationships between the above have been developed. The causal factors and its mitigation measures are taken from published literature as well as from industry practice and experts' opinion. The relationships embedded in the CPTs capture the probabilistic nature of the relationships.

Event trees for post LOC and its equivalent BN are also developed separately for Floating and Cone roof tanks. For Floating roof, rim seal fire and spill on the roof and fire due to LOC have been given separately. For Cone roof, three BNs are given, one for post LOC for gas side and another for liquid side. Additionally, a BN for a confirmed gas LOC is given to illustrate how the probability of fire increases on a confirmed gas leak. The models serve as a baseline for the risk levels in atmospheric storage tank installation. Whenever actual data about the site becomes available, it can be entered into the model to see the impact of the same on the probability of LOC.

BN for compressor damage is discussed in the next chapter.

6

The Jaipur Tank Farm Accident

6.1 What Happened at IOC Jaipur Tank Farm: Predictability of Bayesian Network

Retrospective Application of Bayesian Network to Hazard Assessment of a Floating Roof Tank:

Without going into the details, in a concise summary, what happened at Indian Oil Corporation (IOC) Jaipur tank farm was a prolonged loss of containment (LOC) of Gasoline without any control action to mitigate it, for a time longer than normally expected – about more than an hour. The secondary containment as well as the firefighting system failed. In fact, Gasoline spread through storm water drains also. Therefore, even if the operator wanted to do something, nothing was possible. All that could have been done was to pump and spread foam onto the top of this massive pool from a foam truck (if available). However, the vapor cloud already started forming.

Meanwhile about 700,000–800,000 kg of Gasoline spread as a massive pool. It did not catch fire immediately. A vapor cloud progressively formed from the evaporation of the pool and reached about 8800–10,000 kg (estimated) when it found an ignition source that resulted in an unconfined vapor cloud explosion (UVCE) (Lal Committee, 2011).

As can be seen, these event sequences cannot be predicted or modeled beforehand (a priori). However, 'What if' scenarios with Bayesian network (BN) can provide some insight into possible accident paths. For example if we simulate the post-release BN equivalent for Event Tree (ET), with the following:

- LOC = 100%
- Long duration of release greater than 20 min
- Large pool = 100%
- Failure of operator action = 100%
- Failure of firefighting (FF) system = 100%

Then it is seen that the probability of a Vapor Cloud Explosion (VCE) goes from an average of 9.62 e-12 (From Figure 6.1) to 4.8 e-6 (Figure 6.2 Last Node -LateIgnAndVCE) which is a hypothetical case of Jaipur with Operator action and 0.0095 (Figure 6.3, Last Node – LateIgnAndVCE – with failure of operator action & fire fighting system.), which is an indication that the risk of VCE is very high. Ignition probabilities are from IOGP (2010). Details are given in Figure 6.3.

Such type of analysis is generally difficult in traditional Quantitative Risk Assessment (QRA).

Several industrial QRAs as well as those from the internet have been examined, but none of them have described the scenarios that are possible with a well-modeled BN, as noted above.

The following presents a case study for hazard assessment of a Floating Roof tank using BN. In the subsequent sections, the risk profile of the tank at the pre-accident situation of IOC Jaipur fire is described. This is done to see the predictability of the model.

The methodology adopted is to see if the BN for Floating roof tank described in Chapter 5 can give a degree of prediction when the parameters existing before the accident, i.e. pre-accident conditions, are input to the model's parent nodes. As a quick check, the parameters of all the parent nodes were changed to 'Poor' to see the result of BN simulation in a worst-case situation. The result will give an idea of the probability of LOC when all conditions are in worst state.

The BN model for a Floating roof tank described in Chapter 5, Section 5.3, contains the probability values in a normal 'Good' situation when the facility is assumed to be designed, constructed, operated & maintained with compliance to all codes and standards, complying mostly to all the required systems & procedures, average quality of risk assessments and average score in the quality of Human & Organizational Factors. The values for the node states in 'Good' state is reproduced in the fourth column in Table 6.1. The values are from BN, Figures 5.5–5.11 in Chapter 5.

Next step involved revising the probability values of the parent nodes to have a predominantly 'Poor' state for the main causal factors. When the BN is simulated with these values a risk picture of what can be 'a worst case' emerges. The fifth column titled 'Poor' state values contain results from such values. It can be seen that for all the key causal factors, the state of 'Poor' is very high, and consequently the probability of LOC is also very high.

To access the realistic situation at the Jaipur facility before the accident, the last step is to revise the probability values of the parent nodes of the same BN model based on the pre-accident conditions similar to those existing at IOC Jaipur. The pre-accident condition probabilities are assigned based on the Lal committee investigation committee report (Lal Committee, 2011). BN simulated with the site situation produced values of the key casual factors listed in the last column of Table 6.1.

TABLE 6.1

BN Values for Good & Poor and Pre-Accident Conditions Before the Accident

	Node Name	Node States	'Good' State Values. Probability % (From Figures 5.5–5.11)	'Poor' State Values. Probability %	Values Before Accident. Probability %
1	Quality of design	Poor	0.023	99.9	43.0
		Average	5.97	0.10	50.7
		Good	94.0	0.0	6.29
2	Quality of Maint. & inspection	Poor	0	99.9	99.7
		Average	0. 72	0.061	0.27
		Good	99.3	0.0	0.0
3	Quality of construction	Poor	0.043	99.9	45.2
		Average	8.0	0.56	35.7
		Good	92.0	0.0	19.1
4	Quality of equipment selection	Poor	0.064	99.9	81.6
		Average	7.79	0.082	17.8
		Good	92.2	0.0	0.6
5	Quality of risk assessments	Poor	21.3	70.5	92.6
		Average	35.5	24.2	7.40
		Good	43.2	5.34	0.044
6	Quality of systems & procedures	Poor	39.1	100.0	100.0
		Average	46.2	0.044	0.044
		Good	14.7	0.0	0.0
7	Quality of organizational factors	Poor	98.9	99.8	99.8
		Average	1.06	0.16	0.16
		Good	0.0	0.0	0.0
	Failure probability shell		4.86 e-06	2.0 e-05	1.0 e-05
	Failure probability liquid side.		1.80 e-06	3.2 e-03	2.9 e-03

6.2 Summary of the Investigation Committee Findings

The Lal committee investigation committee report notes the critical factors that resulted in the accident as

i. Loss of primary containment of Motor Spirit (Petrol)

ii. Loss of secondary containment

iii. Incapacitated Operating Personnel

iv. Inadequate mitigation measures

    v. Shortcomings in design and engineering specifications of facilities and equipment

    vi. Absence of Operating Personnel from site and also from vital operational area

Root cause parameters in BN were changed based on the above findings. The values calculated by BN for these conditions (pre-accident situation) is given in the last column of Table 6.1, which basically gives an idea about the risk situation existing in the facility during that time.

From Table 6.1, it can be seen that the facility was operating very near to the 'Poor' state, meaning that the probability LOC was very high when compared with the normal state of such type of facilities.

6.3 BN for Post LOC ET

It is interesting to see the BN simulated values for a post LOC on liquid side, that is, the ET, for a Floating Roof tank. The generic probability values from industry references have been used in the BN shown in Figure 6.1.

Three cases have been simulated in BN and given in Figures 6.1–6.3:

1. With probability of LOC values available from references, namely 2.0 e-06 (Please see the first node – LOCPipeLiqOrSmall of Figure 6.1).
   The probability of late ignition and VCE (last node-LateIgnAndVCE) is only 9.6187 e-12.

2. A hypothetical case with a confirmed LOC from pipe or shell, long duration, large pool, detection and with operator action and with no failure of firefighting system (Figure 6.2, BN for post LOC Floating roof tank – Jaipur scenario – Long Duration – with Operator action & FF System
   The probability of late ignition and VCE (last node-LateIgnAndVCE) now increases to 4.7981 e-06.

3. With a confirmed LOC from pipe or shell, long duration, large pool, no detection and no operator action and with failure of firefighting system (Figure 6.3, BN for post LOC Floating roof tank – Jaipur scenario – Long Duration – Failure of Operator action & FF System

The probability of late ignition and VCE (last node-LateIgnAndVCE) now increases to 0.009545, which indicates a high probability.

The BN for post LOC (Figure 6.1) from pipe or shell shows that the probability of fire inside the bund is relatively quite small due to the safeguards

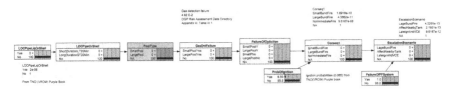

FIGURE 6.1
BN for post LOC of floating roof tank – generic probability of bund fire.

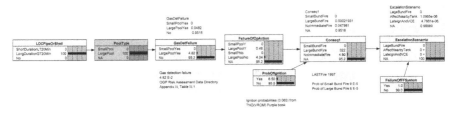

FIGURE 6.2
BN for post LOC Floating roof tank – Jaipur scenario – Long Duration – with Operator action & FF System

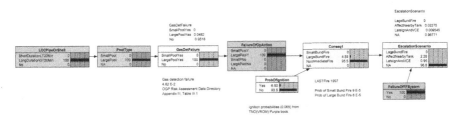

FIGURE 6.3
BN for post LOC Floating roof tank – Jaipur scenario – Long Duration – Failure of Operator action & FF System

preventing an escalation of a fire. The probability values are from the LASTFIRE report (LASTFIRE, 2001) [55].

The second scenario is almost similar to the Jaipur accident. The assumptions include a confirmed LOC (100%) as shown in Figure 6.2 for a longer time, formation of a large pool, no detection but with operator action and with availability of firefighting system.

The BN in Figure 6.3 tries to simulate the case of what actually happened at Jaipur. There is prolonged release liquid, formation of a large pool, delayed ignition (which results in massive evaporation of volatile compounds) and potential for VCE, together with failure of operator action non-availability of firefighting system.

Figure 6.3 presents a worst-case scenario. The BN predicts a high probability of 0.009545 for a VCE, when there is prolonged LOC, together with large pool fire plus failure of operator action and failure of firefighting system.

Traditional QRA methods, though consider sensitivity of failure probabilities, do not normally include such predictions with multiple coincident occurrences and its consequences, which is what actually happens in an accident.

6.4 Chapter Summary

This chapter's main objectives were to show how BN can be applied to model and simulate several cases, including scenarios when critical systems can fail coincidentally. The Jaipur tank farm accident was chosen because details of the investigation committee report were available. By changing the parameters of the nodes of BN suitably, the precursor status at the Jaipur tank farm can be represented to a certain measure. Simulation of the various situations, including the coincidental failures at the tank farm, brings out the high probability of an accident at the site, as compared with published failure data. Such simulations are normally not done during traditional QRA.

7

Bayesian Network for Centrifugal Compressor Damage

Centrifugal compressor is one of the most commonly used rotating equipment in oil and gas industry. The function of the compressor is to increase the pressure of the gas from inlet to outlet. A general schematic of the protective barriers in the compressor system is depicted in Figure 7.1. The automatic emergency shutdown and safety blowdown system is triggered when predefined abnormal conditions are reached by a set of process variables.

7.1 Compressor Failure Modes

The Offshore and Onshore Reliability Data OREDA (OREDA, 2002) lists compressor failure modes as per Table 7.1.

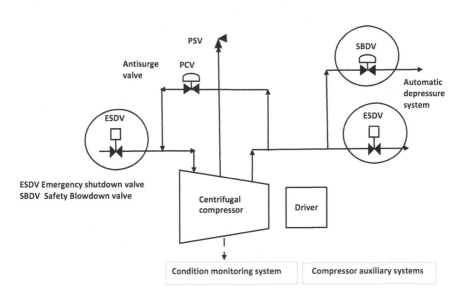

FIGURE 7.1
Simplified schematic diagram of compressor layers of protection.

TABLE 7.1

Failure Modes for a Centrifugal Compressor

Sl. No.	Failure Modes
1	Abnormal instrument reading
2	Breakdown
3	Erratic output
4	External leakage – Process medium
5	External leakage – Utility medium
6	Fails to start on demand
7	Fails to stop on demand
8	High output
9	Internal leakage
10	Low output
11	Minor in-service problems
12	Noise
13	Other
14	Overheating
15	Parameter deviation
16	Spurious trip
17	Structural deficiency
18	Unknown
19	Vibration

7.2 Compressor Failure Rates

As seen from Table 7.1, external leakage of process medium will only result in a loss of containment (LOC) for which (OREDA, 2002) has given a mean failure rate of 10.26 per 10^{-6} hours or 8.99×10^{-2} failures per compressor year. On the other hand, HSE UK reports vide Item FR 3.1.3 (Health and Safety Executive, 2012), frequency rates for rupture as 2.9×10^{-6} per compressor year and 2.7×10^{-4} for small holes 25–75 mm. It is worthwhile to note that one of the main causes for compressor failure in the industry, that is liquid carryover, is not mentioned by either databases but is clearly highlighted by consultant Barringer (Barringer, 2012). It could be that such invisible root cause is not reported properly to the databases.

For the purpose of this study the following causes and sub-causes have been used in the Bayesian network (BN). Please see the simplified schematic for compressor layers of protection.

Expert opinion has been elicited for developing the influencing factors for compressor damage. These causes and sub-causes along with the mitigating factors are mapped into BN. The table of causal factors and the corresponding BN for compressor damage is given in Table 7.2 and Figure 7.2, respectively.

TABLE 7.2

Main and Sub-Causal Factors for Compressor Damage

Sl. No.	Main Causal Factor	Sub-Causal factors. (Parent Nodes)	Notes: All Sub-Causal Factors (Parents Nodes) are Modeled as Normal Distribution with Parameters Noted Below. Ranges of States 'Yes' & 'No' Are Also Indicated
1	Overpressure in compressor	1.1 Downstream blockage 1.2 Failure of valve to flare 1.3 Failure of high pressure shutdown of Emergency Shut Down Valve (ESDV)	Mean = 0.08, SD = 0.1, Yes (0.01–0.025), No (0.025–0.99) Mean = 0.15, SD = 0.1, Yes (0.01–0.06), No (0.06–0.99) Mean = 0.04, SD = 0.01, Yes (0.001–0.002), No (0.002–0.99)
2	Failure of anti-surge valve		Mean = 1.16 e-4, SD = 0.0001, Yes (1e-6-3e-6), No (3e-6-0.001)
3	Intermediate2	3.1 Pressure Safety Valve (PSV) Failure 3.2 PSV undersized	Mean = 0.005, SD = 0.001, Yes (0.001–0.0012), No (0.0012–0.99) Mean = 0.05, SD = 0.1, Yes (0.01–0.02, No (0.02–0.99)
4	Liquid carryover	4.1 Suction demister design inadequate 4.2 Liquid slugs in inlet gas 4.3 Failure of control system 4.4 Failure of high liquid level trip 4.5 Failure of operator action	Mean = 0.15, SD = 0.01, Yes (0.01–0.06, No (0.06–0.99) Mean = 0.15, SD = 0.1, Yes (0.01–0.06), No (0.06–0.99) Mean = 0.12, SD = 0.1, Yes (0.01–0.045, No (0.045–0.99) Mean = 0.09, SD = 0.01, Yes (0.01–0.025), No (0.025–0.99) Mean = 0.11, SD = 0.1, Yes (0.01–0.04, No (0.04–0.99)
5	Gas seal failure	5.1 Liquid carryover 5.2 Excessive vibration	Mean = avg(Sub-causes 4.1–4.5), SD = 0.1, Yes (0.01–0.3), No (0.03–0.99) Mean = 0.0004, SD = 0.01, Yes (1e-4–8e-4), No (8e-4-0.99)
6	Lube oil system failure	6.1 Failure of lube oil system 6.2 Failure of operator action	Mean = 0.006, SD = 0.1, Yes (0.001–0.008), No (0.008–0.99) Mean = 0.11, SD = 0.1, Yes (0.01–0.04, No (0.04–0.99)
7	Foreign object entry	7.1 Temporary strainers 7.2 Lack of procedure for removal/verification	Manual entry, Yes (0.75), No (0.25) Manual entry, Yes (0.10), No (0.90)
8	Change of operating conditions		Manual input. Binary (Yes or No)
9	Thrust bearing failure		Manual input. Binary (Yes or No)
10	Coincident mechanical & acoustic frequencies		Manual input. Binary (Yes or No)

As seen above all the respective sub-causes have been combined in each of the main causal factors using Normal distribution with suitable mean, Standard Deviation (SD) and ranges, except the factors Foreign object entry, Change of Operating conditions, Thrust bearing failure and Coincident mechanical & acoustic frequencies which have manual input.

Example of combining of a main causal factor is given in Equation 7.1

P (LiquidCarryOver | SuctionSDemDesignInAdequate, LiquidSlugsInInlet,

FailureOfControlSys, FailureOfHLvlTrip, FailureOfOpAction1) =

NormalDist (LiquidCarryOver, avg (SuctionSDemDesignInAdequate,

LiquidSlugsInInlet, FailureOfControlSys, FailureOfHLvlTrip,

FailureOfOpAction1), 0.1)   (7.1)

A combination of all causal factors for compressor damage is modeled using NoisyOr distribution and given in Equation 7.2:

P (CompressorDamage | ForeignObjectEntry, LubeOilSyFailure,

FOfCompressorSafetySDSystem, FOfASValveScore, LiqCOScore, GSFScore) =

NoisyOrDist (CompressorDamage, 0.001, ForeignObjectEntry, 0.20,

LubeOilSyFailure, 0.10, FOfCompressorSafetySDSystem, 0.15,

FOfASValveScore, 0.05, LiqCOScore, 0.30, GSFScore, 0.20)   (7.2)

These causal factors and its mitigation measures have been modeled as a BN in Figure 7.2.

7.3 Findings from the BN for Compressor Damage

The BN in Figure 7.2 represents cause and effect relationships of the variables involved in the damage scenarios of a centrifugal compressor. Various predictive and diagnostics mode can be simulated on the BN. For the given set of probability values shown in Figure 7.2, the probability of compressor damage is calculated as 0.0699.

Now if we suspect that the suction scrubber and demister design are inadequate, the corresponding node 'SuctionSDemDesignInAdequate' can be made 100% and the probability of liquid carryover and compressor damage goes up to 22.3 and 10.2 from 13.1 and 6.99, respectively. Please see Figure 7.3.

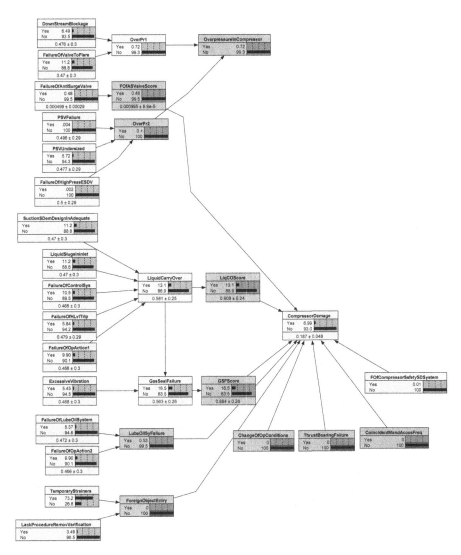

FIGURE 7.2
BN diagram for compressor damage.

Inadequate design and presence of liquid slugs at the inlet will increase the probability of liquid carryover and compressor damage still higher to 38.2 and 15.8, which in relative scale (that is almost a three times increase in probability of liquid carryover and two times increase in compressor damage) is not an acceptable situation. See Figure 7.4.

The BN can be run on diagnostic mode. Given the prior probabilities, it is observed that there is liquid carryover and we want to know which is the most contributing factor. The node 'LiquidCarryOver' is made 100%. BN computes the probabilities of the root causes backwards. This is shown in

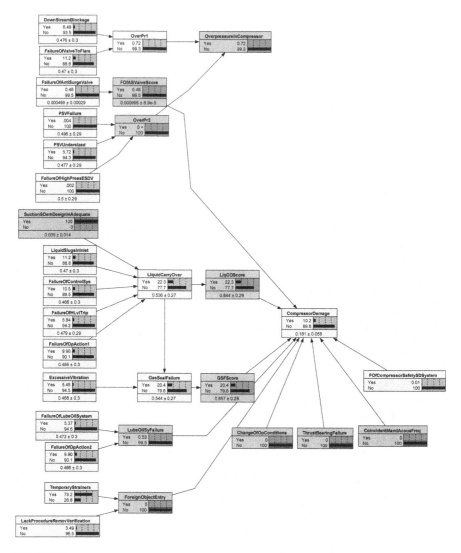

FIGURE 7.3

BN for compressor damage – suction scrubber design inadequate.

Figure 7.5. It is seen that the probability of failures of the parent nodes as well as other nodes have gone up.

The main contributors to the effect 'LiquidCarryOver', as can be ascertained from the BN, in the decreasing order of probability are 'Failure of operator action', 'Liquid slugs at inlet', 'Failure of control system', 'Suction scrubber demister design inadequate', and 'Failure of high level trip'. Such diagnostic mode simulations are valuable tools to understand the root causes of several abnormal situations in a process system.

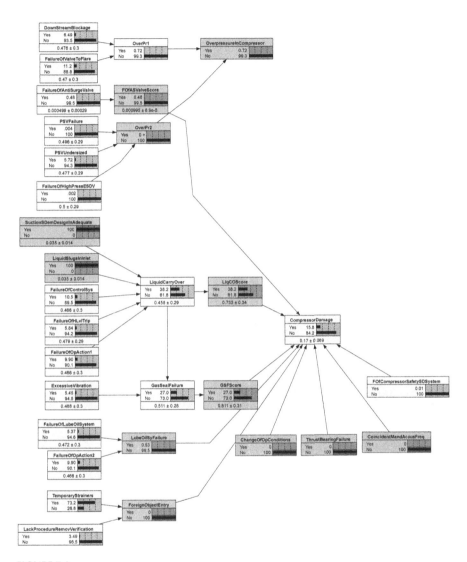

FIGURE 7.4
BN for compressor damage – suction scrubber design inadequate and liquid slugs present.

7.4 Sensitivity of Compressor Damage Node to Parent Nodes

To summarize the impact of parent nodes on the effect node 'CompressorDamage' a sensitivity analysis was performed. Results are shown in Figure 7.6. The analysis results indicate that the failure of operator action followed by liquid slugs at the inlet have the greatest influence on the probability of compressor damage.

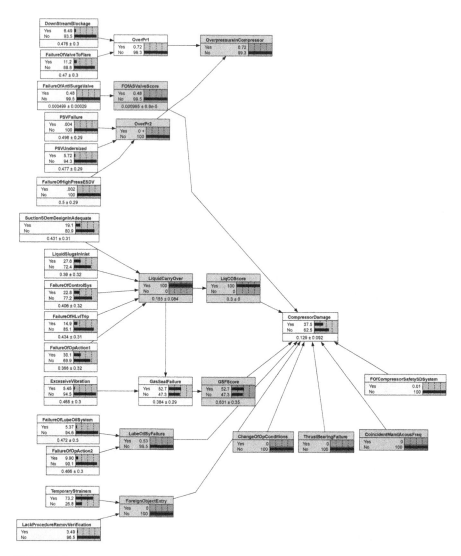

FIGURE 7.5
BN for compressor damage – diagnostic mode – liquid carryover 100.

7.5 LOC and Its Consequences

The LOC from the compressor is usually taken as failure of associated piping and leak from the machine itself, which is very rare.

For average/high reactive gas release, the probability of immediate ignition is 0.2 for release rates below 10 kg s, 0.5 for release rates between 10

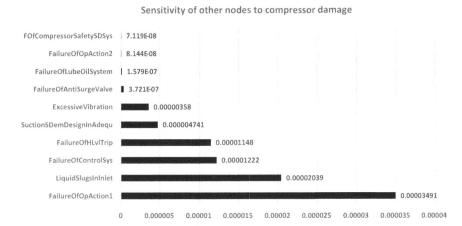

FIGURE 7.6

Sensitivity of other nodes to compressor damage.

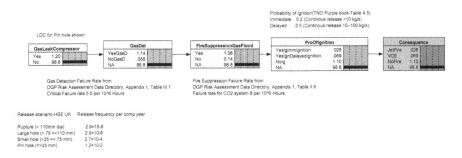

FIGURE 7.7

BN for post LOC-leak from compressor.

and 100 kg s and 0.7 for higher than 100 kg s. Delayed ignition is supposed to occur if the flammable cloud meets an onsite ignition source or if it crosses the site boundary (TNO (VROM), 2005). The BN for LOC for compressor is given in Figure 7.7. A pinhole release rate is taken to illustrate the scenarios.

7.6 Chapter Summary

The BN for compressor damage is described in this chapter. The main causes that result in major damages of the compressor have been ascertained from industry reports and expert opinion. Sub-causes or root causes for these main causes are further listed. Their inter-relationships are given in the table and converted to BN. The use of BN for predictive and diagnostic reasoning is described. Further 'Sensitivity of findings' from Netica indicates that the presence of liquid slugs at the compressor inlet and failure of operator action constitutes the most possible cause for a major damage. The Event Tree and its equivalent BN are also shown.

8

Bayesian Network for Loss of Containment from a Centrifugal Pump

8.1 Introduction

Centrifugal pumps are the workhorses of the process industries that are generally reliable. Apart from damages and minor leaks, a full-fledged loss of containment (LOC) is not that frequent. However, there are components in a centrifugal pump that are susceptible for failures, which can trigger an LOC. One of the more common of such failures is that of the mechanical seal.

Mechanical seal ensures that the pump does not leak and serves to seal the fluid inside the casing from leaking through the opening for the shaft that is rotating. Please see Figure 8.1 for a generic schematic of mechanical seal.

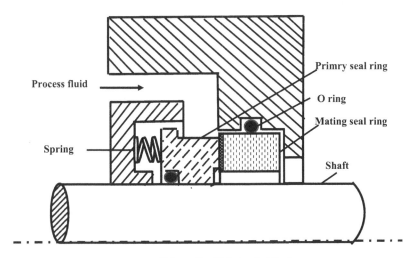

Schematic of Mechanical Seal

FIGURE 8.1
Schematic diagram of a mechanical seal.

8.2 Causes of LOC a Centrifugal Pump

Given below is a list of the more common causes for LOC in a centrifugal pump.

   i. Failure of mechanical seal
  ii. Failure of casing
 iii. Failure of gasket at pump suction/discharge

The causes behind the failure of each of the above are explained below in detail.

8.2.1 Mechanical Seal

   i. Letting the pump or machine run dry: If the pump is running dry, there is a chance for increased heating up and damage of the mechanical seal.
  ii. Excessive vibration: Pump vibration can happen due to improper alignment, pump imbalance or operation beyond the operating limits, and the above can damage the seal.
 iii. Hammering in the associated piping components: Such hammering can happen in certain other connected process equipment, transmitted to the pump, and can cause seal damage.
  iv. Human error: Improper installation, operation of pump and lack/improper maintenance are the potential causes for seal failure.
   v. Use of wrong seal: It is a possibility that the wrong type of seal gets installed, which can cause seal damage eventually.

8.2.2 Casing

   i. Wrong metallurgy
  ii. Change of process fluid causing corrosion
 iii. Casting defects

8.2.3 Suction or Discharge Gasket/s

   i. Wrong selection of gasket
  ii. Wrong installation

The contribution of each of the above towards an LOC is illustrated in Table 8.1. These are based on industry reports and feedback from experts.

TABLE 8.1

Main and Sub-Causal Factors for LOC in a Centrifugal Pump

Sl. No.	Main Causal Factor (Weightage %)	Sub-Causal Factors. Parent Nodes (Weightage%)	Notes: All Sub-Causal Factors (Parents Nodes) are Binary States with 'Yes' & 'No' with Probability Values (Manual Input). Main Causal Factors are Modeled with NoisyOr Distribution with Weightage Factors)
1	Mechanical seal failure (90)	1.1 Pump runs dry (65) 1.2 Excessive vibration (5) 1.3 Hammering on piping components (5) 1.4 Particulates in seal fluid (5) 1.5 Error in installation (10) 1.6 Wrong seal (10)	P (MechanicalSealFailure | ErrorInInstallation, ExcessiveVibration, HammeringPipingComponents, ParticulatesInSealFluid, PumpRunsDry, WrongSeal) = NoisyOrDist (MechanicalSealFailure, 0.0001, PumpRunsDry, 0.65, ExcessiveVibration, 0.05, HammeringPipingComponents, 0.05, ParticulatesInSealFluid, 0.05, WrongSeal, 0.10, ErrorInInstallation, 0.10)
2	Casing failure (2)	2.1 Wrong metallurgy (15) 2.2 Change of Process fluid (77) 2.3 Casting defects (5) 2.4 Excessive cavitation (3)	P (CasingFailure | WrongMetallurgy, ChangeOfProcessFluid, CastingDefects, ExessiveCavitation) = NoisyOrDist (CasingFailure, 0.00001, WrongMetallurgy, 0.15, ChangeOfProcessFluid, 0.77, CastingDefects, 0.05, ExessiveCavitation, 0.03)
3	Gasket failure at suction and discharge flanges (8)	3.1 Wrong gasket 3.2 Wrong installation	P (GasketFailureAtSDFlanges | WrongGasket, WrongInstallation) = NoisyOrDist (GasketFailureAtSDFlanges, 0.000001, WrongGasket, 0.60, WrongInstallation, 0.40)

The failure rates for the Mechanical seal, Casing and Gaskets are of the order of 10^{-4}, 10^{-5} and 10^{-6}, respectively. They are from Failure rates & event rates given by Health and Safety Executive (Health and Safety Executive, 2012). Failure rates for all parent nodes are not readily available from published literature. Expert opinion and industry reports have been used for the same and are readily modifiable as and when actual data is available for an installation.

There are several other causes, for example bearing high temperature, lube oil failure, etc., for a pump damage or failure, but they will not result in an LOC from the pump as such. Therefore, such causes are not included in the modeling of LOC for a centrifugal pump.

8.3 BN for LOC in a Centrifugal Pump

The main and sub-causal factors of LOC have been captured in a Bayesian network (BN), and the same is given in Figure 8.2.

Sensitivity of the root causes (parent nodes) are given below in two figures. Figure 8.3a gives the overall sensitivity. However, the contribution of the most common root cause of LOC, the dry running of the pump, is way too high to fit the scale for other root causes. Therefore, the sensitivity of other causes is shown in Figure 8.3b.

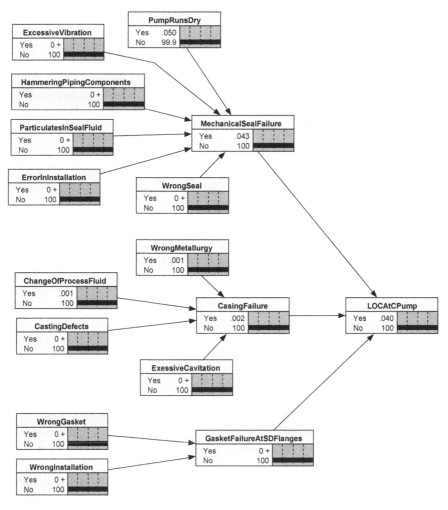

FIGURE 8.2
BN for LOC of a centrifugal pump.

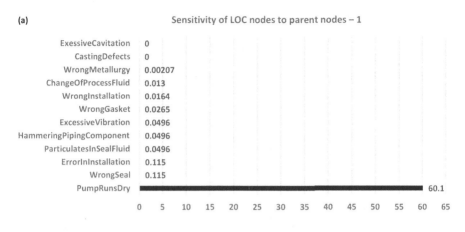

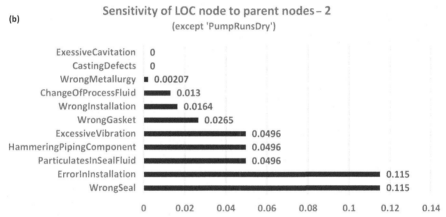

FIGURE 8.3
Sensitivity of LOC node to (a) parent nodes-1, (b) parent nodes-2 (excepting PumpsRunDry).

8.3.1 Consequences of LOC from a Centrifugal Pump

The consequences of LOC are represented in an ET and the same is given in Figure 8.4.

Apart from the 'Yes' & 'No' branching at the ET junctions, the figure indicates the abbreviations in brackets that are used in BN for representing the ET.

The combined BN showing both the causes and consequences are given in Figure 8.5. As can be seen, the probability of LOC for a centrifugal pump is generally very low – of the order of 10^{-4}. However, consequences of the same can be sometimes disastrous.

Figure 8.5 is the BN for a simple forward simulation of the consequences using the industry average failure rate of approximately $4 * 10^{-4}$ per pump per year. The BN can be run in diagnostic mode with a finding of a small pool fire. Please see BN in Figure 8.6 with the small pool fire set to 100.

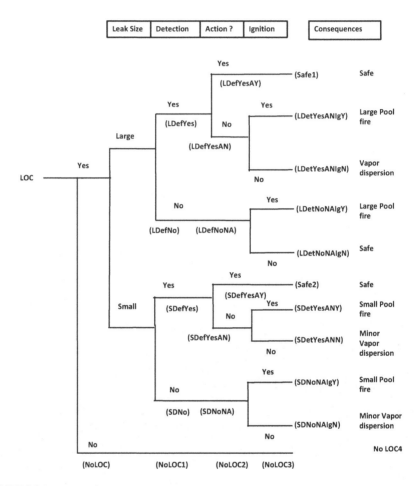

FIGURE 8.4
ET for post LOC for a centrifugal pump.

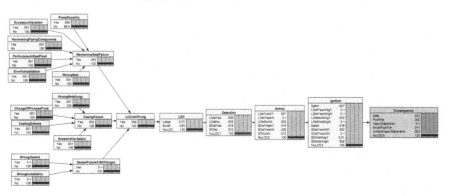

FIGURE 8.5
BN for LOC of a centrifugal pump and its consequences.

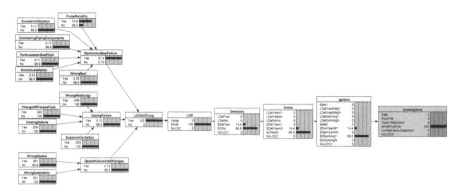

FIGURE 8.6
BN for diagnostic mode – post LOC of centrifugal pump – small pool fire 100.

As expected, the BN predicts the most probable primary cause (0.973) as 'MechanicalSealFailure' with the highest probability of root causes (parent nodes) being 'PumpRunsDry' (0.738) followed by 'Errorininstallation' (0.0023) & 'WrongSeal'(0.0023).

8.4 Chapter Summary

The chapter presented first an analysis of the possible causes for an LOC in a centrifugal pump and second the consequences of such an LOC. Mechanical seal failure has been identified as the most immediate cause of LOC. The causes and its root causes as well as the consequences have been converted to basic BN and simulated. The simulations show how various cases, both causal and diagnostic, can be easily analyzed. The BN can provide a live and visual representation of risks during the operation of a centrifugal pump.

9

Other Related Topics

9.1 Introduction

Application of Bayesian techniques to a hazardous situation will not be complete without a brief introduction to Bayesian inference. Basically, Bayesian inference is a technique used to update a hypothesis about a situation when evidence becomes available. Section 9.2 gives a summary of the technique. Interested readers can explore the subject further with the references given.

Section 9.3 gives a comparison between Quantitative Risk Assessment (QRA) and Bayesian network (BN). QRA has its place in land use planning, but BN can be used to augment the already established QRA technique. Further it needs to find more applications in the oil and gas industry to appreciate its power and flexibility.

9.2 Bayesian Inference

The following is in continuation to what is mentioned in Chapter 2 under 2.1.3, Bayes Formula for Conditional Probability. Suppose we believe that the probability of failure of a Basic Process Control System is 1 in 10 years, or 0.1/year, what do we do when actual data comes in from the facility that there is one failure in the first 2 years and two failures in the next 3 years. Can we predict what will be the failure rates in future, given that we have a prior belief and actual data for 5 years as above.

The above can be put in a formal equation

$$P(H_i \mid E) = P(H_i) * \frac{P(E \mid H_i)}{P(E)} \tag{9.1}$$

where $P(H_i)$ is the probability of H_i (H_i being our belief or hypothesis), $P(E)$ is the evidence expressed as probability.

The equation states that if the right-hand side that is, the prior probability $P(H_i)$ and conditional probability of observing E if H_i, is true, and the total probability of E P(E) is known, then we can compute the left-hand side, which is an update of our hypothesis called posterior probability $H_i|E$. The above can be visualized in a Venn diagram given in Figure 9.1

The hypothesis $P(H_i)$ can be discrete or continuous.

9.2.1 Computational Aspects

When continuous distributions are used the equation takes the form

$$\pi 1(\theta|x) = \frac{p(x|\theta)\,\pi(\theta)}{\int f(x|\theta)\,\pi(\theta)d\theta} \qquad (9.2)$$

where θ is an unknown parameter, for which a prior belief is assigned in probability density function (pdf) as $\pi(\theta)$, before we see any data or evidence.

Next a statistical model called likelihood function is chosen to represent the beliefs about x given θ, described as $p(x|\theta)$.

After observing evidence about x, the update or posterior distribution is computed as $\pi 1(\theta|x)$.

The challenge is in computing the denominator of equation, namely the integral $\int f(x|\theta)\,\pi(\theta)d\theta$.

The computation of the above integral can be avoided, and the problem is transformed to a simple algebra by making the likelihood function a conjugate pair of prior distribution. Conjugate pairs are distributions from the same family and makes the computations easy.

An example is given below for a better understanding of the updating mechanism based on the book from NASA that is available in the public domain Dezfuli, Kely, Smit, Vedros, and Galyean (2009) and Unnikrishnan, Siddiqui and Srihari (2015a).

Let us assume (believe) that the failure of Pressure Safety Valve (PSV) is best described by beta distribution. Beta distribution is non-negative, has a

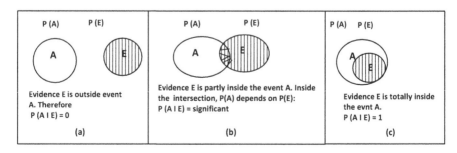

FIGURE 9.1
Venn diagrams representing the situation of Bayesian inference.

range 0–1 and can represent a wide range of situations by adjustment of its shape factors α and β. Conceptually α is the number of failures and $\alpha + \beta$ is the number of demands during which the failure occurs.

a. Taking beta distribution for prior
Beta distribution for PSV is parameterized as α prior = 1.5 and β prior = 200.

$$\text{Mean of Beta distribution} = \frac{\alpha}{(\alpha + \beta)} = \frac{1.5}{(1.5 + 200)} = 0.007444$$

Beta distribution can be readily graphed with the above information using common spreadsheets like Excel. The curve C1 in Figure 9.2 represents the prior distribution.

b. Posterior No. 1
Now we have the data about failure when there has been a demand. One failure of the PSV to open is 100 demands. Binomial distribution is suitable for such a situation such as the failure mechanism for PSV, when there is n number of trails (demands) and k number of failures. Taking binomial distribution, we have the advantage that the posterior will also be a beta distribution since it is the conjugate of binomial.
The α posterior and β posterior can be readily computed from the relationship:

$$\alpha \text{ posterior1} = \alpha \text{ prior} + \text{Number of failures}$$

$$= 1.5 + 1 = 2.5$$

$$\beta \text{ posterior1} = \beta \text{ prior} + \text{Number of demands} - 2$$

$$= 200 + 100 - 2 = 298$$

$$\text{Mean posterior1} = \frac{\alpha \text{ posterior1}}{(\alpha \text{ posterior1} + \beta \text{ posterior1})} = \frac{2.5}{(2.5 + 298)} = 0.008319$$

The curve C2 in Figure 9.2 represents the above posterior distribution. Please note that the mean has shifted towards the right.

c. Posterior No. 2
During the next period, it has been reported that there have been two failures in 100 demands. Here we take posterior1 as prior in the equation. Repeating the steps under b, with the prior as posterior1:

$$\alpha \text{ posterior2 } = \alpha \text{ prior}(\text{posterior1}) + \text{Number of failures}$$

$$= 2.5 + 2 = 4.5$$

$$\beta \text{ posterior2 } = \beta \text{ prior}(\text{posterior1}) + \text{Number of demands} - 2$$

$$= 298 + 100 - 2 = 396$$

$$\text{Mean posterior2} = \frac{\alpha \text{ posterior2}}{(\alpha \text{ posterior2} + \beta \text{ posterior2})} = \frac{4.5}{(4.5 + 298)} = 0.01487$$

The curve C3 in Figure 9.2 represents the above posterior distribution.

As can be seen the uncertainty in the prior distribution has been reduced by the observed data. However, when conjugate pairs cannot be used, advanced techniques such as Markov Chain Monte Carlo (MCMC) with specialized software (WinBugs) are needed.

9.3 Comparison between Traditional QRA and BN Methods

Traditional QRA used for risk assessment has its own strengths and weaknesses. However, decisions makers must understand its limitations. BN methodologies are powerful tools that need to be used more and more so that its strengths and weaknesses can be known and studied. Table 9.1 gives a comprehensive comparison of traditional QRA.

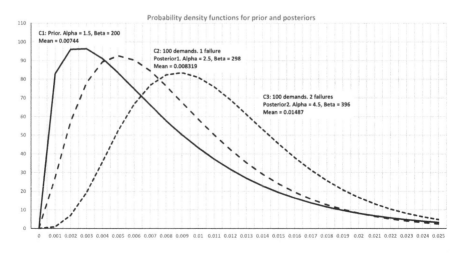

FIGURE 9.2
Probability density function for prior and posteriors.

TABLE 9.1

Comparison Between Traditional QRA and BN Methods

Sl. No.	Parameter	QRA	BN Approach	Notes
1	Scenario (Hazard) identification	Identified generally as LOC from leaks of various hole sizes in pipes and failures of vessel or tank.	BN can include LOC due to leaks from holes in piping, failures of vessels & tanks plus any other credible scenarios including failure of Human and Organizational Factors (HOF)	
2	Failure (LOC) frequencies	Usually taken from published sources (See Table 2.5). Site specific data is not available in most of the cases.	In the BN, initial failure frequencies are taken from published sources. These frequencies are updatable easily to include data and observations at site.	
3	Causal factors for hazard scenarios (LOC)	Once credible scenarios are finalized, causal factors are not considered for analysis.	Causal factors are considered, including non-technical factors like HOF. Cause & effect mechanisms are the most important aspects of BN. Intermediate causes as well as root causes can be modeled in BN.	Frequencies of occurrence of causal factors are included as fixed values or more realistically as probability functions in BN. For example, NoisyOr distribution can describe the effect of many causes far better than a fixed number.
4	Failure (LOC) frequency update for a specific facility	Not usually available. The calculations will have to be repeated with another set – without much basis – which requires time and effort.	Because BN includes causal factors for intermediate causes and root causes, failure frequencies can be updated realistically based on the probability of occurrence of the causal factors.	The probabilities for root causes are combined in a probabilistic manner using Boolean logic or other suitable probability distribution (see Section 2.4).
5	Common cause failures	Almost never considered	Can be considered easily based on the BN logic.	BN basically represents a Bow-Tie diagram, with the left side Fault Tree diagram and the right side Event Tree diagram with the LOC event in the middle. Therefore, examining the same for common causes failure is easy.

(Continued)

TABLE 9.1 (*Continued*)

Comparison Between Traditional QRA and BN Methods

Sl. No.	Parameter	QRA	BN Approach	Notes
6	Expert opinion	Not usually included	Can be suitably included.	Expert opinion can be included using suitable parameters.
7	Safety barriers	Not included explicitly. Credit for the safety barriers is taken in the form of factors to modify the failure frequencies in a deterministic manner.	Modeled explicitly. Barrier failures and its frequencies are part of the Fault & Event Tree mapped into BN. Causes of barrier failures can be included. Please see Kujath, Amyotte, and Khan (2010).	Deterioration of safety barriers (which is usually the cause for escalation of minor accidents) can be reflected in BN by using appropriate equations.
8	Modifications in a facility and its impact	Addition of an equipment or system is usual during operational phase & could affect the risk profile, but this will require revising the QRA, almost never done in practice.	Changes to an equipment or system can be included in BN and its effect can be analyzed.	For example adding a reliable gas detection system will provide a safety barrier, which can be easily added in BN.
9	Including other hazard scenarios	When QRA is done for a total facility, generally other scenarios that have hazard potential like Lightning strike or overfill (in the case of tanks) are generally not considered, unless specifically identified during hazard identification. (Generally only hazards thought to be having the highest possibility are analyzed).	BNs can be developed for such specific cases with all relevant cause and effect mechanisms that can provide a quick and easy assessment of a risk situation. This is possible because BN can analyze all effects and its causes in a visually clear manner.	
10	Finding the most likely causes for an event	Not possible within the QRA framework	BN is a model of all cause and effect relationships with probability values. Therefore it can be run in diagnostic mode to see which nodes (parents) are the highest contributors to an event node (child) that is selected in the BN.	Diagnostic mode is a powerful tool to visualize which is having the largest influence in causing an accident.

(*Continued*) |

TABLE 9.1 (*Continued*)

Comparison Between Traditional QRA and BN Methods

Sl. No.	Parameter	QRA	BN Approach	Notes
11	Transparency of the model	Not very transparent. Specialist assumptions are not always stated.	The model is transparent, visually appealing and the cause and effect mechanism is easily understandable. Experts knowledge can be captured and placed in the model	People who have knowledge about the system can quickly appreciate and learn the model's cause and effect relationship and fine-tune it to represent the real situation.
12	Application area	Most of the QRAs in the industry are oriented to the spatial aspects of risk assessment, that is, for Land Use Planning or for specifying safe spacing criteria.	BN can be used for the same and also for understanding specific risks as noted above.	If the BN for a system is fine-tuned and kept up to date, it can be used for understanding the risk profile of a system at any time.
13	Consistency of result	Wide variations are possible from different analysts for the same system. Example is the ASSURANCE project. (Lauridsen, Kozine, and Markert, 2002)	Since the model is transparent and all assumptions are known, variations are minimum from different analysts.	
14	Use during operational phase of a facility	QRAs done during design stage are usually not available during the operational phase of a facility.	BN model is a live model that can take data on near misses or accident precursors during the operational life of a plant	Risk profile changes during operational period of a facility due to various reasons. To reflect this is very difficult in QRA but possible in BN.
15	Sensitivities	Sensitivity of the results to failure frequency, ignition probability, spillage area, population distribution and vulnerability criteria can be investigated in QRA in a deterministic manner by redoing the calculation with lower and upper bound values for the selected failure frequencies. But the basis for such values are questionable.	In addition to the sensitivity calculations (possible under QRA), BN can compute realistically with sound technical basis, the sensitivity of all causal factor nodes (parent nodes) to an effect node (child node) very easily.	Such analysis enables priorities to be assigned mitigating specific causal factors and in maintenance and testing of safety barriers.

References

Abimbola, M., Khan, F., & Khakzad, N. (2014). Dynamic safety risk analysis of offshore drilling. *Journal of Loss Prevention in the Process Industries, 30,* 74–85.

Ale, B., Gulijk, C. V., Hanea, A., Hanea, D., Hudson, P., Lin, P.-H., & Sillem, S. (2014). Towards BBN based risk modelling of process plants. *Safety Science, 69,* 48–56. doi: 10.1016/j.ssci.2013.12.007.

ANU Enterprise-The Murray Darling Basin Authority. (2016). *Developing a Bayesian Network for Basin Water Resource Risk Assessment-Technical Report.* Retrieved from http://www.mdba.gov.au/kid/files/733-Bayesian-risk-assessment-iCAM.pdf.

Argyropoulos, C., Christolis, M., Nivolianitou, Z., & Markatos, N. (2012). A hazards assessment methodology for large liquid hydrocarbon fuel tanks. *Journal of Loss Prevention in the Process Industries, 25*(2), 329–335.

Atherton, W., and Ash, J.W. (2005). *Review of Failures, Causes & Consequences in the Bulk Storage Industry.* National Lightning Safety Institute. Retrieved from http://www.lightningsafety.com/nlsi_lls/Causes-of-Failures-in-Bulk-Storage.pdf.

Barringer, P. (2012). *Compressors and Silent Root Causes of Failure.* Retrieved from https://silo.tips/download/compressors-and-silent-root-causes-for-failure.

Bayraktarli, Y., Ulfkjaer, J., Yazgan, U., & Faber, M.H. (2005, June 19–23). On the application of Bayesian probabilistic networks for earthquake risk management. In Augusti, G., Schueller, G.I., Ciampoli, M. *9th International Conference on Structural Safety and Reliability (ICOSSAR 05),* Rome, Millpress.

Bearfield, G., & Marsh, W. (2005). Generalizing Event Trees Using Bayesian Networks with a Case Study of Train Derailment. In: Winther, R., Gran, B.A., Dahll, G. (Eds.), Computer Safety, Reliability, and Security. SAFECOMP 2005. *Lecture Notes in Computer Science,* vol 3688. Berlin, Heidelberg: Springer

Bobbio, A., Portinale, L., Minichino, M., & Ciancamerla, E. (2001). Improving the analysis of dependable systems by mapping fault trees into Bayesian networks. *Reliability Engineering & System Safety, 71*(3), 249–260.

Buncefield Major Incident Investigation Board. (2007). *Recommendations on the Design and Operation of Fuel Storage Tanks.* Retrieved from http://www.hse.gov.uk/comah/buncefield/miib.

Buncefield Major Incident Investigation Board. (2008). *The Buncefield Incident 11 December 2005-The Final Report.* Retrieved from http://www.hse.gov.uk/comah/buncefield/miib.

Cai, B., Liu, Y., Zhang, Y., Fan, Q., Liu, Z., & Tian, X. (2013). A dynamic Bayesian networks modeling of human factors on offshore blowouts. *Journal of Loss Prevention in the Process Industries, 26*(4), 639–649.

Center for Chemical Process Safety. (2001). *Guidelines for Chemical Process Quantitative Risk Analysis.* New York: Wiley-AIChE.

Chang, J. I., & Lin, C.-C. (2006). A study of storage tank accidents. *Journal of Loss Prevention in the Process Industries, 19*(1), 51–59.

Davis, P.M., Diaz, J-M., Gambardella, F., & Uhlig, F. (2013). *Performance of European Cross Country Oil Pipelines, Report No 12/13. Statistical Summary of Reported Spillages in 2012 and Since 1971.* CONCAWE, Brussels. Retrieved from http://www.concawe.org.

Det Norske Veritas (DNV). (2013). *Failure Frequency Guidance. Det Norske Veritas (DNV).* Retrieved from www.dsb.no/Global/10%20Vedlegg%20D.pdf.

Dezfuli, H., Kely, D., Smith, C., Vedros, C., & Galyean, W. (2009). *Bayesian Inference for NASA Probabilistic Risk and Reliability Analysis, NASA SP-2009-569.* National Aeronautics and Space Administration. Retrieved from https://ntrs.nasa.gov/archive/nasa/casi.ntrs.nasa.gov/20090023159.pdf.

Duan, R.-X., & Zhou, H.-L. (2012). A new fault diagnosis method based on fault tree and Bayesian networks. *Energy Procedia, 17,* 1376–1382.

E & P Forum. (1992). *Hydrocarbon Leak and Ignition Database,* Report No. 11.4/180. London.

European Gas Pipeline Incident Data Group. (2014). *9th Report of the European Gas Pipeline Incident Data Group (period 1970-2013).* European Gas Pipeline Incident Data Group (EGIG). Netherlands. Retrieved from www.EGIG.eu.

Fenton, N.E., & Neil, M. (2012). *Risk Assessment and Decision Analysis with Bayesian Networks.* Boca Raton, FL: Taylor & Francis.

Flemish Government. (2009). *Handbook of Failure Frequencies 2009 for Drawing up a Safety Report.* Belgium: LNE department. Environment, Nature and Energy Policy Unit, Safety Reporting Division.

Gulvanessian, H., & Holický, M. (2001). Determination of actions due to fire: Recent developments in Bayesian risk assessment of structures under fire. *Progress in Structural Engineering and Materials, 3*(4), 346–352.

Health and Safety Executive. (2012). *Failure Rate and Event Data for use within Risk Assessments, 2012.* Health and Safety Executive, UK. Retrieved from http://www.hse.gov.uk/landuseplanning/failure-rates.pdf.

Health and Safety Executive. (2016). *An Assessment of Measures in Use for Gas Pipelines to Mitigate Against Damage Caused by Third Party Activity,* Contract Research Report 372/2001. Health and Safety Executive, UK. Retrieved from http://www.hse.gov.uk/research/crr_pdf/2001/crr01372.pdf.

Herbert, I. (2010). The UK Buncefield incident – The view from a UK risk assessment engineer. *Journal of Loss Prevention in the Process Industries, 23*(6), 913–920.

Huang, S.-Y., & Mannan, M. S. (2013). Technical aspects of storage tank loss prevention. *Process Safety Progress, 32*(1), 28–36.

Hubbard, D. W. (2009). *The Failure of Risk Management.* Hoboken, NJ: John Wiley & Sons Inc.

INERIS. (2004). *Institut National de l'Environnement Industriel et des Risques. Accidental Risk Assessment Methodology for Industries-ARAMIS User Guide.* Retrieved from http://safetybarriermanager.duijm.dk/files/aramis/ARAMIS_FINAL_USER_GUIDE.pdf.

International Association of Oil & Gas Producers. (2010). *Process Release Frequencies-Report No. 434-1.* Retrieved from http://old.ogp.org.uk/pubs/434-01.pdf.

International Association of Oil & Gas Producers. (2010). *Ignition Probabilities-Report No. 434-6.1.* Retrieved from http://www.ogp.org.uk/pubs/434-06.pdf.

International Association of Oil & Gas Producers. (2010). *Storage Incident Frequencies-Report No. 434-03.* Retrieved from http://www.ogp.org.uk/pubs/434-03.pdf.

Kalantarnia, M., Khan, F., & Hawboldt, K. (2010). Modelling of BP Texas city refinery accident using dynamic risk assessment approach. *Process Safety and Environmental Protection, 88*(3), 191–199.

Kang, J., Liang, W., Zhang, L., Lu, Z., Liu, D., Yin, W., & Zhang, G. (2014). A new risk evaluation method for oil storage tank zones based on the theory of two types of hazards. *Journal of Loss Prevention in the Process Industries, 29,* 267–276.

Khakzad, N. (2012). Dynamic Safety Analysis Using Advanced Approaches, PhD [Dissertation]. St. James, New Foundland: Faculty of Engineering and Applied Sciences, Memorial University of New Foundland. http://research.library.mun.ca/view/departments/FacEngineering.html.

Khakzad, N., Khan, F., & Amyotte, P. (2011). Safety analysis in process facilities: Comparison of fault tree and Bayesian network approaches. *Reliability Engineering & System Safety, 96*(8), 925–932.

Khakzad, N., Khan, F., & Amyotte, P. (2013a). Dynamic safety analysis of process systems by mapping bow-tie into Bayesian network. *Process Safety and Environmental Protection, 91*(1–2), 46–53.

Khakzad, N., Khan, F., & Amyotte, P. (2013b). Quantitative risk analysis of offshore drilling operations: A Bayesian approach. *Safety Science, 57,* 108–117.

Khakzad, N., Khan, F., & Amyotte, P. (2013c). Risk-based design of process systems using discrete-time Bayesian networks. *Reliability Engineering & System Safety, 109,* 5–17.

Kjaerulff, U. B., & Madson, A. E. (2008). *Bayesian Networks and Influence Diagrams: A Guide to Construction and Analysis.* New York: Springer.

Korb, K. B., & Nicholson, A. E. (2010). *Bayesian Artificial Intelligence,* 2nd edition. London: Chapman & Hall/CRC Computer Science & Data Analysis.

Krishnan, V. (2006). *Probability and Random Processes.* Hoboken, NJ: John Wiley & Sons Inc.

Kujath, M., Amyotte, P., & Khan, F. (2010). A conceptual offshore oil and gas process accident model. *Journal of Loss Prevention in the Process Industries, 23*(2), 323–330.

Lal, MB Committee. (2011). *Report of the Committee on Jaipur Incident.* Oil India Safety Directorate. Retrieved from http://www.oisd.gov.in.

Lannoy., A. & Cojazzi., G.M. (Eds.), (2002, Nov., 18–19). *23rd ESReDA Seminar. Decision Analysis: Methodology and Applications for Safety of Transportation and Process Industries: Using Influence Diagrams to Analyze and Predict Failures in Safety Critical Systems,* Delft, The Netherlands. Retrieved from http://www.bookshop.europa.eu/.../decision-analysis/LBNA21004ENC_00.

LASTFIRE. (2001). *BP Safety Series, Large Atmospheric Storage Tank Fires.* Rugby, UK: IChemE.

Lauridsen, K., Kozine, I., & Markert, F. (2002). *Assessment of Uncertainties in Risk Analysis of Chemical Establishments-The ASSURANCE Project,* Risø National Laboratory, Roskilde, Denmark. Retrieved from https://backend.orbit.dtu.dk/ws/files/7712279/ris_r_1344.pdf.

Marsh & McLennan. (2011). *Atmospheric Storage Tank-Risk Engineering Position Paper-01,* Marsh & McLennan, UK. Retrieved from https://usa.marsh.com /uk/insights/research/risk-engineering-position-paper-atmospheric-storage-tanks-html.

McGrayne, S. B. (2011). *The Theory that Would not Die.* New Haven, CT: Yale University Press.

Montgomery, D. C., & Runger, G. C. (5th ed.). (2011). *Applied Statistics and Probability for Engineers.* Hoboken, NJ: John Wiley & Sons Inc.

Muhlbauer, K. (2004). *Pipeline Risk Management Manual, Third Edition: Ideas, Techniques, and Resources.* Cambridge, MA: Elsevier.

Necci, A., Argenti, F., Landucci, G., & Cozzani, V. (2014). Accident scenarios triggered by lightning strike on atmospheric storage tanks. *Reliability Engineering & System Safety, 127*, 30–46.

Netica. (2017). *Vancouver: Norsys Software Corporation.* www.norysys.com.

Ni, Z., Phillips, L. D., & Hanna, G. B. (2011). Exploring Bayesian Belief Networks Using Netica®. In Darzi, A., & Athanasiou, T. (Eds.), Evidence Synthesis in Healthcare-A Practical Handbook for Clinicians (pp. 293–318). London: Springer-Verlag London.

OREDA. (2002). *Offshore Reliability Data,* 4th ed. OREDA Participants. Available from: Det Norske Veritas, NO-1322, Hovik, Norway

Pasman, H. J., & Rogers, W. J. (2012). Risk assessment by means of Bayesian networks: A comparative study of compressed and liquefied H_2 transportation and tank station risks. *International Journal of Hydrogen Energy, 37*(22), 17415–17425.

Pasman, H., & Rogers, W. (2013). Bayesian networks make LOPA more effective, QRA more transparent and flexible, and thus safety more definable! *Journal of Loss Prevention in the Process Industries, 26*(3), 434–442.

Pasman, H., Jung, S., Prem, K., Rogers, W., & Yang, X. (2009). Is risk analysis a useful tool for improving process safety? *Journal of Loss Prevention in the Process Industries, 22*(6), 769–777.

Pettitt, G., & Morgan, B. (2009). A tool to estimate the failure rates of cross country pipelines, *Symposium Series, Hazards XX1 Process Safety and Environmental Protection in a Changing World, IChemE Symposium Series No. 155,* November 10–12, Manchester, UK. Retrieved from https://pdfs.semanticscholar.org/e6a3/508b6aedac0bc5f72f25ce01310021228664.pdf.

Pouret, O., Naim, P., & Marcot, B. (Eds.). *Bayesian Network: A Practical Guide to Applications.* New York: Wiley-AIChE.

Press Trust of India, (2014, Sept. 10). GAIL pipeline fire due to collective failure: Oil ministry probe report, *The Economic Times.*

Ramnath, V. (2013). Key aspects of design and operational safety in offsite storage terminals. *Hydrocarbon Processing, 92*, March, 71–73.

Rathnayaka, S., Khan, F., & Amyotte, P. (2012). Accident modeling approach for safety assessment in an LNG processing facility. *Journal of Loss Prevention in the Process Industries, 25*(2), 414–423.

Roy, A., Srivastava, P., & Sinha, S. (2014). Risk and reliability assessment in chemical process industries using Bayesian methods. *Reviews in Chemical Engineering, 30*(5) 479–499.

Stover, R. (2013). *Review of the US Department of Transportation Report: The State of National Pipeline Infrastructure.* Retrieved February 10, 2016 from http://www.icog.com/~oildrop/PHMSA report analysis.pdfitate.

Straub, D. (2005, June 19–23). Natural hazards risk assessment using Bayesian networks. In Augusti, G., Schueller, G.I., Ciampoli, M. *9th International Conference on Structural, Safety and Reliability (ICOSSAR 05),* Rome: Millpress.

Tan, Q., Chen, G., Zhang, L., Fu, J., & Li, Z. (2014). Dynamic accident modeling for high-sulfur natural gas gathering station. *Process Safety and Environmental Protection, 92*(6), 565–576.

TNO (VROM). (2005). *Guidelines for Quantitative Risk Assessment, Purple Book, CPR 18E*. GEVAARLIJKE STOFFEN. Retrieved from https://content.publicatier-eeksgevaarlijkestoffen.nl/documents/PGS3/PGS3-1999-v0.1-quantitative-risk-assessment.pdf.

Tuft, P. (2014). *Comparing International Pipeline Failure Rates*. The Australian Pipeliner, 124–134, Retrieved from http://pipeliner.com.au/news/comparing_international_pipeline_failure_rates/86968.

Twardy, C. R., Nicholson, A. E., & Korb, K. (2005). *Knowledge Engineering from Cardiovascular Bayesian Networks from the Literature*. Technical Report TR 2005/170, Clayton School of IT, Monash University.

Tweeddale, M. (2003). *Managing Risk and Reliability in Process Plants*. Cambridge, MA: Elsevier.

Unnikrishnan, G., Shrihari, H., & Siddiqui, N. (2014). *Application of Bayesian Methods to Event Trees with case Studies, Reliability Theory and Applications, #03 (34)*, Vol.9.

Unnikrishnan, G., Shrihari, H., & Siddiqui, N. (2015a). Monitoring probability of failure on demand of safety instrumented systems by Bayesian updating. *International Journal of Applied Engineering. Research*, 10, 35774–3577.

Unnikrishnan, G., Shrihari, H., & Siddiqui, N. (2015b). Understanding oil & gas pipeline and mitigation measures using Bayesian approach. *International Journal of Applied Engineering. Research*, 10, 29595–29608.

Weber, P., Medina-Oliva, G., Simon, C., & Iung, B. (2012). Overview on Bayesian networks applications for dependability, risk analysis and maintenance areas. *Engineering Applications of Artificial Intelligence*, 25(4), 671–682.

Index

Printed in the United States
by Baker & Taylor Publisher Services